SpringerBriefs in Fire

Series editor:

James A. Milke
Department of Fire Protection Engineering
University of Maryland
College Park, Maryland, USA

More information about this series at http://www.springer.com/series/10476

Matthew E. Benfer • Joseph L. Scheffey

Evaluation of Fire Flow Methodologies

 Springer

Matthew E. Benfer
Hughes Associates, Inc.
Baltimore, MD, USA

Joseph L. Scheffey
Hughes Associates, Inc.
Baltimore, MD, USA

ISSN 2193-6595   ISSN 2193-6609 (electronic)
SpringerBriefs in Fire
ISBN 978-1-4939-2888-0 ISBN 978-1-4939-2889-7 (eBook)
DOI 10.1007/978-1-4939-2889-7

Library of Congress Control Number: 2015944710

Springer New York Heidelberg Dordrecht London

Printed on acid-free paper

Springer Science+Business Media LLC New York is part of Springer Science+Business Media (www.springer.com)

Foreword

The basic method for controlling building fires by fire departments is through the use of water, which is typically applied with manual hose lines or water monitors. This water can come from a municipal water supply, a private water supply, or from the fire department itself (i.e., water tenders). In order to effectively fight a fire, the water supply available must be adequate for the threat from the building and contents. The water requirements for firefighting include the rate of flow, the residual pressure required at that flow, the flow duration, and the total quantity of water required. As described in the *NFPA Handbook* [1], the American Water Works Association (AWWA) [2] defines the required fire flow as "the rate of water flow, at a residual pressure of 20 psi and for a specified duration that is necessary to control a major fire in a specific structure." Each fire flow methodology may define the objective of the required fire flow differently. There are a number of methods currently used to calculate required water flow rates for sprinklered and non-sprinklered properties. These methods are, in general, based on decades-old criteria derived using data from actual fires. Over the years, building construction methods, building contents, and fire suppression equipment and tactics have changed. The overall objective of this study is to assess the appropriateness of currently available fire flow methodologies. The first task in this project was a literature review of the existing fire flow calculation methodologies in the USA and globally. The second task was a data analysis and GAP assessment to determine what additional information is needed to validate the existing fire flow calculation methodologies.

References

1. Wenzel, L.J., "Water Supply Requirements for Public Supply Systems," *NFPA Handbook, 20th Edition,* Section 15, Chapter 2, Quincy, MA.
2. AWWA M31, Distribution *System Requirements for Fire Protection,* American Water Works Association, Denver, CO, 1989.

Baltimore, MD, USA

Matthew E. Benfer
Joseph L. Scheffey

Acknowledgments

The authors thank the Fire Protection Research Foundation via the National Fire Protection Association for sponsoring the production of this material.

The authors would like to acknowledge the help of Marty Ahrens and the NFPA Fire Analysis Division for providing the NFIRS data used. The authors would also like to acknowledge the input of the technical panel for providing guidance and support during this project, particularly Ronnie Gibson from Factory Mutual and Jeff Shapiro from International Code Consultants.

The members of the technical panel are listed here:

Tony Apfelbeck, Altamonte Springs
Jeff Collins, Palm Beach County
John Devlin, Aon Fire Protection Engineering
Chris Farrell, NFPA Public Fire Protection
Dawn Flancher, American Waterworks Association
Scott Futrell, Futrell Fire Design and Consult
Ronnie Gibson, FM Global
Tom Helgeson, CH2MHill
Bruce Johnson, International Code Council
Kevin Kimmel, Clark Nexen
Jeff Shapiro, International Code Consultants
Will Smith, Code Consultants, Inc.
Greg Harrington, NFPA staff liaison

Contents

Acronyms

AHJ	Authority Having Jurisdiction
AWWA	American Water Works Association
BSI	British Standards Institute
CC	Construction classification number
FEDG	Fire engineering design guide
FF	Fire flow
FIERA	Fire evaluation and risk assessment
FLED	Fire load energy densities
FPRF	Fire Protection Research Foundation
FSRS	Fire Suppression Rating Schedule
FW	Fire water
GPM	Gallons per minute
HRR	Heat release rate
ICC	International Code Council
IFC	International Fire Code
IIT	Illinois Institute of Technology
ISO	Insurance Service Office
ISU	Iowa State University
IWUIC	International Wildland-Urban Interface Code
NBFU	National Board of Fire Underwriters
NFA	National Fire Academy
NFF	Needed fire flow
NFIRS	National Fire Incident Reporting System
NFPA	National Fire Protection Association
NZ	New Zealand
OHC	Occupancy hazard classification number
PAB	Public access building
SFPE	Society of fire protection engineers

UK	United Kingdom
US	United States
WS	Water supply
WTRM	Water requirements model

Chapter 1
Introduction

There are a number of methods currently used to calculate required water flow rates for sprinklered and non-sprinklered buildings. These methods are, in general, based on decades-old criteria derived using data from actual fires. The overall objective of this study was to assess the appropriateness of currently available fire flow methodologies. The first task was a literature review of the existing fire flow calculation methodologies in the US and globally. The second task was a data analysis and GAP assessment to determine what additional information is needed to validate the existing fire flow calculation methodologies.

As part of the literature review, 19 existing fire flow calculation methods were identified. The methods identified are from the US, UK, France, Germany, the Netherlands, New Zealand, and Canada. Two types of methods were evaluated; those for building planning (building and/or fire code requirements); and, those for on-scene fire service use. The building planning methods accounted for a range of variables in determining fire flow. The on-scene fire flow calculation methods consist of one equation with one variable. Other than as a first order approximation, the on-scene methods do not appear to lend themselves for use in codifying requirements. The building planning methods must be predictive. Important variations in building construction, use, or features such as exposure protection, or the installation of sprinklers, are de facto parameters for establishing the magnitude of a fire which an arriving fire department should be able to handle.

Sixteen fire flow calculation equations/methods were evaluated for two different size non-residential buildings and two different size single-family residential buildings. For the residential buildings, only four of the fire flow methods reduced the fire flow with the presence of sprinklers; the reductions were greater than or equal to 33 %. The wide variance in calculated fire flow values is consistent with findings by other researchers and identifies a need for actual field data to validate the methodologies for current building types.

A GAP analysis was conducted to identify the steps needed to determine which methods (and associated variables) result in the most accurate fire flow assessment. The methods used calculate the size of fire expected to be suppressed (and the fire

© Fire Protection Research Foundation 2015
M.E. Benfer, J.L. Scheffey, *Evaluation of Fire Flow Methodologies*,
SpringerBriefs in Fire, DOI 10.1007/978-1-4939-2889-7_1

flow required to do so), whether based on heat release rate or building area/volume. The on-scene methods use simple, scientifically-based formulas to derive the flow based on fire area or volume. However, other metrics are needed to establish the reasonable fire size to be expected. These are embodied in the building planning methods.

Substantial validation data, which was not readily identified in this project, is important to be able to determine the accuracy of fire flow calculation methodologies. Key variables needed include: actual fire flow used as a function of specific building geometry, exposures, and installed protection. The National Fire Incident Reporting System (NFIRS) is widely used by fire departments across the US to report fire incident data. All necessary data to assess fire flow is included in the NFIRS report, except for an explicit estimate of the actual water used to suppress a given fire. The system does not currently contain a specific field for fire flow at an incident. If a fire flow field were added as a required item for NFIRS reporting, the NFIRS fire incident reports would provide validation data for the fire flow methodologies. This is a recommended long term approach.

An attempt was made to obtain fire flow data from public research; this yielded a single result. An FPRF survey of fire departments across the US was conducted to gather data on water usage during residential fire suppression activities. For 25 of the incidents in the FPRF report, the size of the dwelling was able to be determined. Four of the existing fire flow calculation methods were evaluated for comparison to the field data. This comparison showed that the fire flow methodologies were rarely less than the actual fire flows used. This suggests that the fire flow calculations methodologies are conservative, and might be adjusted downward.

More data is required sufficiently validate fire flow methodologies over a range of occupancies. A long-term approach would be to change NFIRS to include reporting of fire flow data. In the near term, a survey of selected fire departments could be performed, similar to the approach used in the FPRF study of residential water meters. Fire departments used in that study could potentially be used again. A possible survey form has been outlined.

Chapter 2
Literature Review

2.1 Existing Flow Methodologies

As part of the literature review, 19 existing fire flow calculation methods have been identified. Sixteen were examined and are described in subsequent sections. Three other foreign fire flow calculation methods were identified, but were not able to be examined due to the lack of a translated copy of the document. The methods identified come from the US, UK, France, Germany, the Netherlands, New Zealand, and Canada. Where methods were very similar, they were grouped together and described in one section. Two types of methods were evaluated including those for building planning (building and/or fire code requirements) and those for on-scene fire service use. The building planning methods accounted for a range of variables in determining fire flow (i.e., building construction, occupancy, fire size, etc.). This allows for building and community planners to assess current or future buildings against the existing or planned water supply and adjust accordingly. The on-scene fire flow calculation methods consist of one equation with one variable used to determine the fire flow. This allows the firefighters on scene to assess whether they need more hose lines or apparatus to fight the fire. All fire flow equations have been converted to imperial units (ft and gal), where appropriate. The 19 methods identified are listed below:

Building Planning Methods

1. ISO Method (US)
2. IFC/NFPA 1 Method (US)
3. NFPA 1142 Method (US)
4. IWUIC Method (US)
5. Ontario Building Code Method (Canada)
6. FIERAsystem Method (Canada)
7. TP 2004/1 and TP 2005/2 Methods (NZ)
8. FEDG Method (NZ)

© Fire Protection Research Foundation 2015
M.E. Benfer, J.L. Scheffey, *Evaluation of Fire Flow Methodologies*,
SpringerBriefs in Fire, DOI 10.1007/978-1-4939-2889-7_2

9. PAS 4509 (NZ)
10. D9 Method (France)
11. UK National Guidance Document on the Provision of Water for Firefighting (UK)

On-Scene Methods

12. ISU Method (US)
13. Särdqvist, Thomas, and Baldwin Methods (UK, UK, and US, respectively)
14. IIT Method (US)
15. NFA Method (US)
16. 3D Firefighting Method (UK/US/Australia)

Other Foreign Methods

17. Dutch Ministry of the Interior Approach *Beheersbaarheid van Brand 2007* (The Netherlands)
18. German VdS 2034 Method (Germany)
19. German DVGW W405 Method (Germany)

2.1.1   ISO Method

The ISO method, contained in ISO PPC3001 [3], has been developed as an aid in estimating the amount of water that should be available for municipal fire protection, otherwise known as the needed fire flow (NFF). ISO uses the needed fire flow at various buildings within a community in order to evaluate the adequacy of the water supply and delivery system for the purpose of establishing insurance premiums. In addition, the needed fire flows within the community are used to determine the firefighting apparatus, size of apparatus fire pumps, and special firefighting equipment needed in the community. As early as 1889, the National Board of Fire Underwriters (NBFU) began to make fire protection surveys of municipalities. The ISO procedure for establishing fire flows evolved from the original NBFU formula, which was used for years to determine the fire flow required in downtown business districts of municipalities. The original NBFU formula grew from a paper by Hutson [4] in 1948. PPC3001 states that ISO developed the needed fire flow calculation method through a review of large-loss fires but does not reference any specific data. Reportedly, the fire flows arose from ISO field engineers counting the number of hoselines used at actual fire scenes and assuming that one 2-1/2 in. hose supplied 250 gpm. This method is only applicable to non-residential, non-sprinklered buildings and residential buildings which are sprinklered or non-sprinklered. The ISO method considers building construction, occupancy, adjacent exposed buildings, and fire communication paths between buildings. The basic ISO formula is:

$$NFF_i = (C_i)(O_i)\left[1 + (X+P)_i\right]$$

where:

NFF_i is the needed fire flow for the subject building (gpm)
C_i is a construction factor that depends on the construction of the building (gpm)
O_i is an occupancy factor that depends on the combustibility of the occupancy
X is a factor related to the exposure buildings
P is a factor related to the communication between buildings

The construction factor, C_i, is calculated from a coefficient, F, based on the class of construction and the effective building area, A_i, using the following formula:

$$C_i = 18F\sqrt[3]{A_i}$$

The construction class coefficient, F, ranges from 0.6 for fire-resistive construction to 1.5 for wood frame construction. Descriptions of the various construction classes are contained in the ISO PPC3001 document with conversions from construction classification types used by other codes (e.g., NFPA and ICC). The effective building area, A_i, is calculated as the total square footage of the largest floor of the building plus a percentage of all other floors based on the ISO construction class of the building. The construction factor, C_i, value is rounded to the nearest 250 gpm. In addition, it shall not be below 500 gpm nor shall it exceed:

- 8000 gpm for Construction Class 1 and 2 (Combustible Construction)
- 6000 gpm for Construction Class 3, 4, 5, and 6 (Non-combustible Construction)
- 6000 gpm for a 1-story building of any construction class

The occupancy factor, O_i, reflects the combustibility of the contents of the building. Occupancy factors range in value from 0.75 for non-combustible (C-1) to 1.25 for rapid burning (C-5) classes. Descriptions of the various occupancy classes (C-1 through C-5) and how buildings with multiple occupancy classes are treated are contained in the ISO PPC3001 document.

The exposure and communication factor, $(X+P)_i$, reflects the need for additional water to mitigate the exposure of adjacent (X) or communicating (P) buildings to fire from the subject building. The maximum value of the exposure and communication factor, $(X+P)_i$, is 0.6. The exposure and communication factor is based on the side of the building with the greatest exposure factor, X, and the greatest communication factor, P, calculated. The exposure factor, X, is dependent up on the construction and length-height value (length of wall in feet, times the height in stories) of the exposure building and the distance between facing walls of the subject building and exposure building. An exposure building is defined as a building with a wall 100 ft or less from a wall of the subject building. Exceptions and tabulated values for the exposure factor, X, and communication factor, P, are contained in the ISO PPC3001 document.

For 1- and 2-family dwellings not exceeding two stories in height, ISO prescribes the following needed fire flows based on the distance between buildings:

- 500 gpm where the distance is more than 100 ft
- 750 gpm where the distance is between 31 and 100 ft

- 1000 gpm where the distance is between 11 and 30 ft
- 1500 gpm where the distance is 10 ft or less.

The ISO method provides for reduction in the needed fire flow when sprinkler systems are installed in the subject building. The sprinkler systems must first be graded using other ISO methods. Where all 1- and 2-family dwellings in an entire subdivision or other definable area are protected with a residential sprinkler system, a reduction in the needed fire flow to 500 gpm at 20 psi is allowed. However, the ISO method does not allow reduction where individual 1- and 2-family dwellings provided with residential sprinkler systems are interspersed with similar non-sprinklered 1- and 2-family dwellings.

ISO filed a revised Fire Suppression Rating Schedule (FSRS) in December 2012 which includes some changes to the NFF calculation [5]. This version is pending approval, but is anticipated to take effect in 2014. The NFF calculation was updated after discussion with local fire authorities and national water, fire, and emergency communication associations, in order to modernize the methodology within the context of the FSRS. The primary changes to the NFF calculation are to the exposure and communication factor tables, but the maximum value of the exposure and communication factor, $(X+P)_i$, remains the same. In addition, the NFF values specified for 1- and 2-family dwellings have been altered as follows:

- 500 gpm where the distance is more than 30 ft
- 750 gpm where the distance is between 21 and 30 ft
- 1000 gpm where the distance is between 11 and 20 ft
- 1500 gpm where the distance is 10 ft or less.

The sprinkler reduction for 1- and 2-family dwellings has been expanded to include all 1- and 2-family dwellings with sprinkler systems regardless of whether they are interspersed with similar non-sprinklered dwellings. In addition, for 1- and 2-family dwellings greater than 4800 ft^2 in area, the NFF calculation methodology outlined in this section is now used rather than the prescribed values for other 1- and 2-family dwellings. The revised FSRS also includes fire flow durations for residential and non-residential buildings.

2.1.2   International Fire Code (IFC) and NFPA 1 Methods

The International Fire Code (IFC) [6] and NFPA 1 [7], the National Fire Protection Association (NFPA) Fire Code, contain very similar methods for determining the required fire flow. Both codes use tabulated values of the needed fire flow which were based on a simplified ISO method. The IFC fire flow calculation procedure is contained in Appendix B of the code and is not a requirement unless specifically adopted by the Authority Having Jurisdiction (AHJ). The procedure in NFPA 1 is a requirement contained in the main body of the code. Both codes intend for the fire flow methodology to be used with new building construction and/or relocated buildings.

The required fire flow for buildings other than one- and two-family dwellings and one- and two-family dwellings greater than 5000 ft^2 (NFPA 1) or 3600 ft^2 (IFC) are tabulated in tables B105.1 and 18.4.5.1.2 for the IFC and NFPA 1, respectively. The fire flows are based on the type of construction of the building as determined by each code and the calculated fire flow area. Fire flows range from 1500 to 8000 gpm at 20 psi residual pressure. Flow durations between 2 and 4 h are required based on the fire flow. For one- and two-family dwellings having a fire flow area less than 5000 ft^2 (NFPA 1) or 3600 ft^2 (IFC), the minimum fire flow is required to be 1000 gpm for 1 h.

In general, the fire flow area is the total floor area of all floor levels of a building. The IFC also accounts for areas under horizontal projections of the roof of a building, while NFPA 1 does not. For Type I (443), Type I (332), and Type II (222) construction in the NFPA 1 regime and Type IA and IB in the IFC regime, the fire flow area is the area of the three largest successive floors. In both codes, portions of buildings separated by firewalls are considered separate buildings for the purpose of calculating required fire flow.

Both the IFC and NFPA 1 allow for increases and decreases to the required fire flow. In the case of unusual susceptibility to group fires or conflagrations, the AHJ can increase the required fire flow by up to 100 %. For isolated buildings or groups of buildings in rural/suburban areas, the AHJ can decrease the required fire flow. Both of these allowances are subjective in nature and are based on the judgment of the AHJ with regards to applicability. Installation of an approved sprinkler system can reduce the fire flow by up to 50 % for one- and two-family dwellings and up to 75 % for all other buildings. In NFPA 1, a 25 % reduction in required fire flow can also be applied when the building is separated from other buildings by a minimum of 30 ft. Both codes also provide minimum required fire flows after decreases have been applied. As is common for model building codes, some local jurisdictions pass amendments to the codes to standardize certain procedures or make other modifications. For instance, the city of San Antonio, TX has added a table to the IFC which specifies the fire flow reduction for sprinklered buildings based on the building construction type, occupancy type, and storage height and arrangement [8].

2.1.3   NFPA 1142 Method

NFPA 1142, *Standard on Water Supplies for Suburban and Rural Fire Fighting* [9], is referenced in the IFC for areas where adequate and reliable water supply systems do not exist. Chapter 4 of this standard outlines the method for calculating the minimum water supply (gal) for basic structural firefighting. NFPA 1142 does not contain the historical basis for the development of the calculation methodology. Annex G of NFPA 1142, which is not part of the requirements of the standard, outlines a method for determining the fire flow required where a municipal-type water system is present. The method contained in Annex G is the same as the ISO method (see Sect. 2.1.1). The required water supply (gal) for basic structural fire-fighting is calculated from Chap. 4 using the following formulas:

$$WS_{min} = \frac{VS_{tot}}{OHC} CC \quad \text{(for structures without exposure hazards)}$$

$$WS_{min} = \frac{VS_{tot}}{OHC} CC * 1.5 \quad \text{(for structures with exposure hazards)}$$

where:

WS_{min} is the required minimum water supply (gal)
VS_{tot} is the total volume of structure (ft^3)
OHC is the occupancy hazard classification number (1–7, 1 is most hazardous)
CC is the construction classification number (0.5–1.5)

Exposure hazards are defined as structures as: 100 ft^2 or larger in area, within 50 ft of the subject building; or, where the structure, regardless of size, has an occupancy classification number of 3 or 4 (see discussion below). The minimum required water supply for structures without exposure hazards is 2000 gal. and the minimum for structures with exposure hazards is 3000 gal. NFPA 1142 allows the AHJ to reduce the water supply require when the building is protected by an approved sprinkler system or other automatic fire suppression system. This code references the fire flow reduction limits used in NFPA 1 (see Sect. 2.1.2) in the explanatory material contained in its appendix. The minimum required water supply can also be increased by the AHJ to compensate for conditions including: limited fire department resources, extended fire department response time, limited access, hazardous vegetation, unusual terrain, special uses, or occupancies, structural attachments such as decks or porches, etc. These increases can be arbitrary and are based on the judgment of the AHJ.

The occupancy hazard classification number is determined based on the use of the building (e.g., plastics processing, schools, etc.). NFPA 1142 lists the different occupancy hazards for each classification number in the body of the code. Where two occupancies are present in a building, the more hazardous occupancy classification number is to be used. Occupancy classification numbers are between 1 and 7. The construction classification number is determined based on the construction type of the building (Type I through Type V). Where more than one construction type is present in a building, the higher construction classification number is to be used. Construction classification numbers are between 0.5 and 1.5.

The minimum fire flow rate required by NFPA 1142 is tabulated, as shown in Table 2.1, based on the calculated minimum water supply. This is the fire flow rate that the fire department is required to have the capability of delivering within 5 min of the arrival of the first apparatus at the incident.

Table 2.1 NFPA 1142 fire
flow rate

Total water supply required (gal)	Fire flow rate (gpm)
<2500	250
2500–9999	500
10,000–19,999	750
≥20,000	1000

2.1.4   IWUIC Method

The International Wildland-Urban Interface Code (IWUIC) [10], part of the International Code Council (ICC) family of codes, has been developed to address the mitigation of fire in wildland-urban interface areas. Wildland-urban interface areas provide unique challenges for firefighting due to their remoteness and challenging fire hazards. The IWUIC, Section 404, regulates both the water source and the minimum fire flow of the water source. Water sources must be approved by the AHJ and can be man-made or natural sources. Approved hydrants are required for access to the water; the access points of the water source are required to be no more than 1000 ft from the building. Access to the water supply must remain unobstructed at all times. In addition, the water sources are subject to periodic testing as required by the AHJ.

The purpose of the water supply, as stated in the code, is to provide for initial structural fire attack and exterior flame front control in the Wildland-urban interface zone. The water flow rate is based on the type and fire flow area (i.e., floor area) of the building. One- and two-family dwellings must have a water supply capable of delivering 1000 gpm for 30 min if the fire flow area is less than 3600 ft^2 or 1500 gpm for 30 min if the fire flow area is greater than 3600 ft. The AHJ is allowed to reduce the fire flow by 50 % when an approved sprinkler system is present. The water supply required for buildings other than one- and two-family dwellings must be approved by the AHJ, but can be no less than 1500 gpm for a duration of 2-hours. It is likely that fire flow for buildings other than one- and two-family dwellings would be calculated based on the IFC requirements (see Sect. 2.1.2) with some modification by the AHJ. The AHJ is allowed to reduce the fire flow by up to 75 % when an approved sprinkler system is present, but the fire flow can be no less than 1500 gpm.

2.1.5   Ontario Building Code Method

The Ontario Building Code [11] provides a method for calculating the required water supply quantity and flow rate for fire fighting in non-sprinklered buildings. For sprinklered buildings, only the hose stream demands and durations required by NFPA 13, Standard for the Installation of Sprinkler Systems [12], are required to be provided. The minimum requirements for water supply quantity are relevant to buildings not serviced by a municipal water supply system. Requirements for buildings serviced by municipal water supply systems, where the water supply duration is not a concern, focus on water supply flow rate (i.e., fire flow) and minimum pressure. The intent of these requirements are to provide a water supply sufficient enough for the fire department to extinguish building fires where adverse circumstances are not encountered [13]. The basic formula for determining the minimum quantity of water required is:

$$Q = (0.00749) * K * V * S_{Tot}$$

where:

Q is the minimum quantity of water (gal)
K is the water supply coefficient
V is the total volume of the building (ft^3)
S_{Tot} is the spatial coefficient to account for exposure protection.

The water supply coefficient, K, is determined from a table contained in the code. These values range from 10 to 53 based on the construction of the building. According to the Ontario Building Code guidelines for fire flow calculation [13], the water supply coefficients were developed from the occupancy hazard classification numbers and construction classification numbers in NFPA 1231 [14] (former name of NFPA 1142), with some adjustment and modification to fit in with the Ontario Building Code. The construction of the building is determined based on other requirements in the Ontario Building code. The spatial coefficient S_{Tot} is the total of the spatial coefficient values from property line exposures on all sides of the building using the following formula:

$$S_{Tot} = 1.0 + [S_{side1} + S_{side2} + S_{side3} + ... S_{sideN}]$$

where, the individual spatial coefficients are determined from a curve contained in the code, with some modifications. This curve accounts for the distance between the building and the exposure building as well as the building occupancy. Individual spatial coefficients are less than or equal to 0.5 and S_{Tot} has a maximum value of 2.0.

The minimum required fire flow rates are tabulated based on the minimum water supply as shown in Table 2.2. For municipal water supplies, the required fire flow rate must be provided for a minimum of 30 min and at a pressure of 20.3 psi.

Table 2.2 Ontario building code fire flow rate

Minimum water supply, Q (gal)	Fire flow rate (gpm)
a	476
≤28,571	714
28,571 < Q ≤ 35,714	952
35,714 < Q ≤ 42,857	1190
42,857 < Q ≤ 50,265	1429
50,265 < Q ≤ 71,429	1667
Q > 71,429	2381

[a]One-story building with building area not exceeding 6456 ft^2

Additions to existing buildings are also required to comply with the minimum water supply requirements. The code recommends that the entire building volume (i.e., existing plus addition) be used to calculate the required water supply and fire flow rate. The maximum fire flow for any water supply is 2381 gpm.

2.1.6   FIERAsystem Water Requirements Model

The Fire Evaluation and Risk Assessment system (FIERAsystem) is a computer model developed by the Canadian National Research Council which is used to evaluate fire protection systems in light industrial buildings [15]. The Water Requirements Model (WTRM) was developed to estimate the water requirements for firefighting purposes. The model considers the geometry of the building, possible fire scenarios, fire detectors, suppression systems, adjacent buildings, and response time and effectiveness of the fire department. Specifically, the model calculates the required fire flow for suppression of the fire and exposure protection at the time of the fire department intervention.

For water suppression estimation, the combined heat release rate (HRR) of all fires in the building at the time the fire department first applies water to the fire is the quantitative basis of required water rate. The FEDG method (see Sect. 2.1.8) of estimating the cooling capacity of firefighting water is used to predict the effect of firefighting on the total HRR. An efficiency factor for water application is applied. See Torvi et al. [15] for additional details regarding this method.

2.1.7   New Zealand SFPE Method TP 2004/1 and TP 2005/2

The New Zealand SFPE published two technical publications relating to water requirements for firefighting purposes. The first, TP 2004/1 [16], presents a calculation method for the water flow requirement for firefighting purposes. The second, TP 2005/2 [17], presents a calculation method for the amount of stored water used for firefighting purposes. The general equation for determining the required fire flow in TP 2004/1 is:

$$F = \frac{k_F * Q_{max}}{k_W * Q_W}$$

where,

F is the required fire flow (gpm)
k_F is the heating efficiency of fire (conservatively 0.5)
k_W is the cooling efficiency of available water (conservatively 0.5 for a water main)
Q_{max} is the maximum heat output of fire (MW)
Q_W is the absorptive capacity of water at 100 °C (0.164 MW/gpm)

If the conservative values of k_F and k_W are used and Q_W is substituted, the equation reduces to:

$$F = 6.1 * Q_{max} \quad (\text{gpm})$$

Further study of floor areas, ventilation opening ratios, and fire load energy densities (FLED) was carried out and an alternative fire flow calculation method was determined based on the FLED and floor area, A_{floor}, of the room or building. The units for FLED are mega-joules (MJ) per unit floor area (i.e., square feet or square meters). This eliminated the Q_{max} variable. The resulting equation is:

$$F = 0.118 * \left(FLED * A_{floor} \right)^{0.666} \quad (\text{gpm})$$

TP 2005/2 recommends a similar method for calculating the minimum volume of water storage, S, needed for firefighting purposed. The equation for S is as follows:

$$S = 0.132 * FLED * A_{floor} \quad (\text{gallons})$$

2.1.8 New Zealand Fire Engineering Design Guide Method (FEDG)

The New Zealand Fire Engineering Design Guide (FEDG) Method [18] is based on principles similar to those used for the New Zealand SFPE methods (see Sect. 2.1.7). The required fire flow is based on the theoretical heat absorbing capacity of water and steam. The following formula is used to calculate the required fire flow, F:

$$F = \frac{Q_F}{\eta_a * Q_W}$$

where,

F is the required fire flow (gpm)
Q_F is the heat release rate of the fire in MW
η_a is the cooling efficiency of the water (conservatively 0.5)
Q_W is the absorptive capacity of water at 100 °C (0.164 MW/gpm)

This formula is only applicable to moderate size fire areas. This is because the burning rate equations developed for use in this method were determined from fires in rooms of about 110 ft².

2.1.9 SNZ PAS 4509 Methods

The New Zealand Fire Service Firefighting Water Supplies Code of Practice, SNZ PAS 4509:2008 [19] was developed to provide direction on what constitutes a sufficient supply of water for firefighting in urban fire districts. This code provides

Table 2.3 PAS 4509 basic required fire flow

Fire water classification	For reticulated water supply		For non-reticulated water supply	
	Required fire flow within 433 ft (gpm)	Additional required fire flow within 866 ft (gpm)	Firefighting time (min)	Minimum water storage within 295 ft (gal)
FW1	119	0	15	1849
FW2	198	198	30	11,888
FW3	397	397	60	47,551
FW4	794	794	90	142,652
FW5	1190	1190	120	285,304
FW6	1587	1587	180	570,607
FW7	Based on Appendices H and J.			

two methods for determining the required fire flow for a particular building. The first step in both methods is to determine the fire water classification number (FW1 through FW7). This number is tabulated based upon the type of building (i.e., single family homes or other structures), sprinkler protection, the fire hazard category (i.e., occupancy) of the building, and the largest fire cell of the building. A fire cell is defined in the New Zealand building code as any space including a group of contiguous spaces on the same or different levels within a building, which is enclosed by any combination of fire separations, external walls, roofs, and floors. For all fire water classifications except FW7, the required water supply is as shown in Table 2.3. For buildings with reticulated water supplies (i.e., municipal water supplies), fire flow requirements are listed. For buildings with non-reticulated water supplies, the minimum water storage volume and firefighting times are listed. Sprinklered structures are either FW1 for single family homes or FW2 for all other sprinklered structures. FW3 through FW7 depend on the hazard and floor area of the largest fire cell in the building. For FW7, Appendices H and J of PAS 4509 must be used to calculate the required fire flow.

Appendix J of PAS 4509 is used to calculate the required fire water supply (i.e., fire flow), M_{tot}, for fire water classification FW7. The basic equation for the required fire water supply is:

$$M_{tot} = M_{water} + M_{exp} \quad (\text{gpm})$$

where, M_{water} is the required water flow for firefighting and M_{exp} is the required water for exposure protection. The following formula is used to calculate the required water flow for firefighting:

$$M_{water} = 9.2 * Q_{max} \quad (\text{gpm})$$

where, Q_{max} is the maximum fire size for the fire cells in the building. Appendix H of PAS 4509 is used to calculate the maximum fire size. The formula for Q_{max} is:

$$Q_{max} = K_1 * K_2 * Q_{fire} \quad (\text{MW})$$

where, K_1 is the human intervention factor, K_2 is the fire safety features factor, and Q_{fire} is the calculated fire size for the fire cell. Values for K_1 are 1.0 for unoccupied buildings or ones without monitoring facilities, 0.9 for occupied buildings where manual firefighting equipment is available, and 0.8 for buildings where occupants are trained in fire fighting operations and are always present. Values for K_2 are 1.0 where no detection system or suppression system is present or a detection system is present without connection to fire service alarm receiving equipment, 0.8 where a detection system with connection to fire service alarm receiving equipment, and 0.1 where a sprinkler system is installed with direct connection to fire service alarm receiving equipment. Q_{fire} is calculated based on the individual fire cells. For each fire cell, the fire size is calculated for the ventilation limited and fuel limited fire scenarios based on equations presented in the document. The maximum of the two values is used. For the fuel limited fire scenario, some information regarding the building use and hazards is required. Calculations for both fire scenarios are presented Appendix H of PAS 4509.

The fire flow required for exposure protection, M_{exp}, is calculated using the following formula:

$$M_{exp} = 1.58 * A_{exp} \quad (gpm)$$

where, A_{exp} is the total surface area of adjacent fire cells and/or structures exposed to a fire cell involved in the fire. An exposed surface is defined as any external cladding on an adjacent structure that is combustible or coated with a combustible coating that can be affected by radiation. Methods for determining whether the surface can be affected by radiation from the fire cell are contained in the New Zealand Building Code.

2.1.10   French D9 Technical Document Method

The French D9 technical document, *External Fire Control—Determination of Water Supply* [20], was prepared by a group of public and private French organizations to act as a guide for determining minimum water requirement for emergency services. The minimum water requirements were based on extinguishing a fire limited to the maximum non-divided surface (i.e., a fire in a portion of the building) and not the entire building. This document applies to homes, offices, high-rise buildings, public access buildings (i.e., commercial buildings), and industrial risks. The method is not intended to be used for tank farms, chemical industries, and other special risks. Separate methods are used for each of the types of buildings, i.e., homes and office buildings including high-rises, public access buildings, and industrial risks.

For homes and office buildings, the minimum fire flows are tabulated based on the height of the building and the non-divided developed surface area. The building height, H, is the maximum of the height of the lowest floor and the highest level with respect to the ground level. The non-divided developed surface, S, is delimited by fire resistant walls and/or floors with 1 h minimum ratings or 2 h minimum ratings for high-rise buildings. Fire flows for offices are between 264 and 1056 gpm; fire flows for residential buildings are between 264 and 528 gpm. The minimum duration for these fire flows is 2 h. For homes and office buildings, no reduction in fire flow is given when sprinkler systems are installed.

For public access buildings (PAB), the minimum fire flows are tabulated based on the occupancy classification and non-divided developed surface, S. For values of S up to 107,600 ft², the fire flows range from 264 to 2576 gpm. For values of S greater than 107,600 ft², the fire flows are to be treated on a case-by-case basis. If the building is protected by a sprinkler system, the required fire flow is the same regardless of the occupancy classification and depends only on S. For values of S up to 322,800 ft², the fire flow values for sprinklered buildings are between 246 and 1585 gpm. The minimum duration for fire flows in PAB is 2 h.

For industrial risks, the D9 document has a more complicated approach for determining fire flow requirements. A simplified version of the water flow equation is the following:

$$Q = 0.024 * S * Risk\ Category * \left[1 + \sum coefficients\right] \quad (gpm)$$

where, S is the reference surface area in square feet, the risk category is dependent on the occupancy and storage arrangement, and the coefficients account for the storage height of products, the building construction, and internal intervention types (e.g., fire brigade or fire alarm system). For non-divided buildings with multiple types of risks (i.e., manufacturing area and storage area), the sum of the individually calculated fire flows is the fire flow for the entire building. For divided buildings with multiple types of risks, the maximum of the individually calculated fire flows is the fire flow for the entire building. The reference surface area is defined as the area of the building which is divided by either 2-hour fire walls or by 33 ft of uncovered, obstruction-free space. The risk category coefficient is between 1.0, 1.5, or 2.0 depending on the occupancy and storage arrangement as outlined in an appendix of the document. The storage coefficient ranges from 0.0 (for storage up to 10 ft) to +0.5 (for storage above 39 ft). For warehouses, without details of the storage height, the storage height is assumed to be the building height minus 3.2 ft. The building construction coefficient is either −0.1 (for fire resistant frames ≥ 1 h), 0.0 (for fire resistant frames ≥ 30 min), or +0.1 (for fire resistant frames < 30 min). The internal intervention coefficients are between −0.3 and −0.1 depending on the type of building monitoring or fire brigade present. For buildings that are sprinkler protected, a reduction in the fire flow requirement of 50 % is applied.

The minimum fire flow rate is 264 gpm; fire flow rates calculated are rounded to the nearest 130 gpm multiple. The minimum duration for fire flows for industrial buildings is 2 h, except in special cases. For industrial risks, the public or private water system can be augmented by water reserves where the fire flow cannot be met. However, the document recommends that at least one third of the required fire flow be provided by the public or private water system for quicker fire attack and to reduce times required to implement the water reserves.

Table 2.4 Minimum fire flow from the UK national guidance document on the provision of water for firefighting

Type of building	Minimum fire flow (for any single hydrant) [gpm]
Housing (semi-detached; ≤ 2 floors)	127
Housing (>2 floors)	317–556
Transportation (garages, service stations)	397
Industry:	
• Up to 1 ha	317
• Up to 2 ha	556
• Up to 3 ha	794
• Over 3 ha	1190
Shopping, offices, recreation, tourism	317–1190
Education, health, community facilities:	
• Village halls	238
• Primary schools/single story health care	317
• Secondary schools/colleges, large healthcare facilities	556

2.1.11   UK National Guidance Document on the Provision of Water for Firefighting Method

The UK National Guidance Document on the Provision of Water for Firefighting [21] was produced by representatives of the water industry, fire service, and governmental agencies. The purpose of this document is to provide a risk assessment methodology for determining water requirements for firefighting; the document is not legally binding. Another objective is to facilitate a liaison between the water industry (supplier) and the fire service (user). Appendix 5 contains the guidelines for determining fire flow requirements. The minimum fire flows are listed for particular types of buildings as shown in Table 2.4. These fire flows are the minimum for a single hydrant and are generally independent of the size of the building. For housing, different flows are required depending on the number of stories present. For industrial applications the flow is dependent on the number of hectares (1 ha = 107,200 ft^2). Different flows are required for small and large healthcare facilities. It is unclear from this document whether the number of hectares for the industrial buildings is the site size or the building size. The document does not give specific guidance regarding the number of fire hydrants.

2.1.12   Iowa State University Method (ISU)

The Iowa State University (ISU) method [22] is a commonly used method based on the amount of water needed to deplete the oxygen in a confined area, when the water is vaporized into steam by the heat of the fire. The required flow in gpm is given as:

$$Required\ Flow = \frac{V}{100}\ (gpm)$$

where V is the enclosed volume (length of the structure multiplied by its width and height) in cubic feet. This method is unique in that it does not consider the occupancy hazard, only the volume of the building to be filled with steam. The ISU method was developed based on hundreds of fire tests conducted by Iowa State University in the 1950s [22].

2.1.13 Särdqvist, Thomas, and Baldwin Methods

Three studies of actual fire flow used during residential and/or non-residential firefighting operations were conducted in the 1950s, 1970s, and 1990s in the UK and the US. Generalizing the results of Thomas [23], Baldwin [24], and Särdqvist [25], the required fire flow, FF, has the following correlation:

$$FF = kA^n\ (gpm)$$

where, k and n are constants based on the individual study and A is the horizontal fire area in square feet. This fire area was likely the entire involved area encountered at the fire scene; these equations do not consider the height of the building. Table 2.5 contains a summary of the fire flow equations from Thomas, Baldwin, and Särdqvist.

2.1.14 Illinois Institute of Technology Method (IIT)

The Illinois Institute of Technology method [1] was based on a survey of 134 fires in the Chicago area. It is unclear what persons specifically developed this method, but given that Baldwin [24] used data from 134 fires in Illinois for his fire flow equation, it is likely that Baldwin's data was the basis for this method. The results of the survey were used with regression analysis to develop fire flow formulas based on building area. The fire flow rate is based on one of the following formulas:

Table 2.5 Fire flow equations from Thomas, Baldwin, and Särdqvist

Researcher	Fire flow equation (gpm)	Location of fires	Number of fires analyzed	Size of fire areas analyzed (ft^2)
Thomas (1959) [23]	$24.2 * A^{0.5}$	UK	48	2150–650,000
Baldwin (1972) [24]	$4.09 * A^{0.66}$	Illinois	134	214–130,000
Särdqvist (1998) [25]	$4.17 * A^{0.57}$	UK	307	Up to 10,720

$$\text{Fire Flow} = 0.00009 * A^2 + 0.5 * A \quad \left(\text{for residential occupancies}\right)$$
$$\text{Fire Flow} = \left(-1.3 \times 10^{-5}\right) * A^2 + 0.42 * A \quad \left(\text{for other occupancies}\right)$$

where A is the area of the fire in square feet. For large, non-residential buildings (> ~32,000 ft²), this method becomes invalid as the fire flow tends to decrease for larger areas due to the negative coefficient on one term in the equation.

2.1.15   National Fire Academy Method (NFA)

In the 1980s, the National Fire Academy (NFA) developed a simple method of calculating fire flow at the scene of a fire [26]. This method was intended to be used by fire fighters at an incident as a tool to aid in determining the amount of water necessary to fight the fire, the apparatus used to deliver the water, and the number of companies required for the incident. This formula was developed by the NFA through a study of a large number of working fires and a survey of fire officers throughout the country. The fire flow formula is given as:

$$\text{Fire Flow} = \frac{L * W}{3} \quad \left(\text{gpm}\right)$$

where L is the length of the involved floor in feet and W is the width of the involved floor in feet. This formula can be expanded to include multiple floors by adding the fire flows for each floor. The NFA suggests that the formula is only reliable if four or fewer floors are involved. This formula can also account for a partially involved floor by multiplying by the percentage involvement of that floor.

2.1.16   3D Firefighting Method

3D Firefighting [27] is a training manual on firefighting techniques and tactics published by the Oklahoma State University Press. This manual was put together by four experienced fire service professionals from the US, the UK, and Australia. The manual includes a fire flow calculation method for use by on-scene fire commanders. The tactical flow rate presented in this manual accounts for fire flow required by indirect fire attack, direct fire attack, and water fog attack methods. The authors proposed a minimum tactical flow-rate (i.e., fire flow) of 0.098 gpm per square foot of compartmental fire involvement. The authors also recommend that this flow rate be increased by at least 50 % where the structural members are involved in the fire. The flow rate proposed in this manual was based on the authors' own research in 1990 and other studies of fire flow.

2.1.17 Other Water Supply Literature

The British Standards Institution (BSI) has drafted Published Document PD 7974-5, *Application of Fire Safety Engineering Principles to the Design of Buildings* [28]. The recommended standards of "fire cover" within the UK Fire Service are addressed. This document states that the fire flow capacity should be related to the size of the building and the risk involved and agreed upon with the approving authority, but does not give a specific calculation method.

The NFPA Handbook provides a good list of the historical data used to develop fire flows. Work from Shedd [29], Fanning [30], Kuichling [31], Freeman [32], and Metcalf [33] is cited. These papers give the details regarding the initial development of American and Canadian waterworks standards at the turn of the twentieth century. Correlations were developed for the number of required hose streams given a city population, but do not include fire flow calculations for individual buildings.

2.2 Other International Approaches

Three other international approaches to calculate fire flow were discovered during this literature search. These methods include the Dutch Ministry of the Interior Approach *Beheersbaarheid van Brand 2007* [34], the German VdS 2034 method [35] for non-public fire departments, and the German DVGW W405 method [36]. Due to the lack of adequate translations of these documents, the purpose of the documents and the exact fire flow calculation methods were unable to be determined.

2.3 Other Reviews of Fire Flow Methodologies

Torvi et al. [15], Särdqvist [37], Barnett [38], and Davis [39] have reviewed many of the fire flow methodologies presented in this literature review; reproductions of data from these authors are presented in Appendix A. Torvi et al. examined the ISO, ISU, IIT, FDEG, and Ontario Building Code methods and compared the results for residential buildings, office buildings, and warehouses of various sizes. The resulting fire flows had large differences, sometimes an order of magnitude difference in the predicted flow rates. They criticized the IIT method for being based on too few (134) fires and only having residential and non-residential divisions. In addition, they noted that the methods evaluated only considered the amount of fire flow required for extinguishment, which for large buildings becomes unrealistic. Presumably, they were referring to the ISU, IIT, and FEDG methods which do not have limits. As a result of this evaluation, they decided to develop a new method for estimating fire flow, the FIREAsystem method [15] as described in Sect. 2.1.6.

Särdqvist [37] first compared the Thomas, Baldwin, and Särdqvist methods. As discussed in Sect. 2.1.13, these three methods are very similar. Särdqvist noted that despite different techniques and conditions, the three methods produce quite similar results. In general, the Särdqvist method produced lower flow rates than the other methods, up to a factor of three difference for larger fire areas. Särdqvist opined that the differences arose from the variations in the fires due to differences in date and geographical location of the fires, changes in building construction and furnishing, and changes in firefighting techniques used over time. Särdqvist then compared the ISO, ISU, and FEDG methods and an approximation of sprinkler flow rates to the other fire flow methods. Sprinkler flow rates were between 14*A and 28.5*A gpm, where A is the area in square feet. Comparing these six methods, there was good agreement for some intervals of fire area but the spread in the data was large. Särdqvist noted that there was no model (ISO, ISU, or FEDG) which gave a good representation of the fire flow data from experimental and fire brigade studies (Thomas, Baldwin, and Särdqvist). Särdqvist also mentioned a fire flow method called Firepro [40] which suggested a water flow rate proportional to the floor area, depending on the ceiling height using the following formulas:

$$Q = 18.5 * A \quad \left(\text{for } 8.2 \, \text{ft tall room} \right)$$
$$Q = 28.2 * A \quad \left(11.8 \, \text{ft tall room} \right)$$

where:

Q is the fire flow (gpm)
A is the floor area (ft^2)

However, there is no discussion by Särdqvist regarding the basis for the Firepro method or where it originated. Further research of this method did not provide any additional information.

Barnett [38] compared the five methods evaluated by Torvi et al. to the TP 2004/1 and TP2005/2, 3D Firefighting, Särdqvist, NFPA 1142, and PAS 4509 methods. It should be noted that Barnett used the 2001 edition of NFPA 1142, which did not have the flow requirements for non-municipal type water supplies. The flow rates calculated by Barnett for NFPA 1142 were from tables similar to the IFC/NFPA1/ISO methods. He calculated fire flows for the ten methods for the same size residential buildings, office buildings, and warehouses that were examined by Torvi et al. Barnett saw very large differences in fire flow values calculated by the different methods, again up to an order of magnitude. Barnett stated that the TP2005/2 method, coincidentally his own, appeared to be the best method because it was based on fire engineering principles and included allowance for exposures.

Davis [39] compared the ISO, ISU, IIT, Thomas, Baldwin, and Särdqvist methods by plotting the fire flow as a function of area between 800 and 22,000 ft^2. For the ISO and ISU methods, he plotted the "high risk" and "low risk" calculated values but it is unclear what the differences in the calculations were. Davis noted that there was a wide variance in flow rate with respect to the area and that a direct comparison of these values is not possible due to the differences in definition of either area or volume. This variance was up to a factor of five difference between the lowest value and the highest value calculated, with the exception of the IIT residential flow rates which grew much quicker for areas greater than 2700 ft^2.

Chapter 3
Fire Flow Methodology Examples and Analysis

Sixteen fire flow calculation equations/methods were evaluated for two different size non-residential buildings and two different size single-family residential buildings. Building areas of 10,000 and 50,000 ft² were used for the non-residential buildings and total floor areas of 1500 and 3500 ft² were used for the residential buildings.

3.1 Non-residential Buildings

For the non-residential buildings, the minimum and maximum possible fire flow rates for each method and each size building were calculated. The range of fire flow values for each method and building size are plotted in Figs. 3.1 (non-sprinklered) and 3.2 (sprinklered) for the 10,000 ft² building and Figs. 3.3 (non-sprinklered) and 3.4 (sprinklered) for the 50,000 ft² building. The following assumptions were made in these calculations:

- Buildings were one non-residential fire area (or fire cell)
- Building height was 20 ft
- For NFPA 1142, a building with a non-municipal water supply was assumed
- The combination of variable values which gave the minimum and maximum flow possible were used
- Fuel-limited fire load values of 0.0232 and 0.0929 MW/ft² were used for all calculations of maximum heat release rate (NZ PAS4509, Appendix E)
- Exposed surface area was one face of a square building of the same size as the subject building

The fire flow methodologies produced ranges of flow rates that were sometimes an order of magnitude different from each other (see Figs. 3.1, 3.2, 3.3, and 3.4). It is expected that the other international approaches not discussed would produce similar differences. The wide variance in fire flow values is consistent with analyses

© Fire Protection Research Foundation 2015
M.E. Benfer, J.L. Scheffey, *Evaluation of Fire Flow Methodologies*,
SpringerBriefs in Fire, DOI 10.1007/978-1-4939-2889-7_3

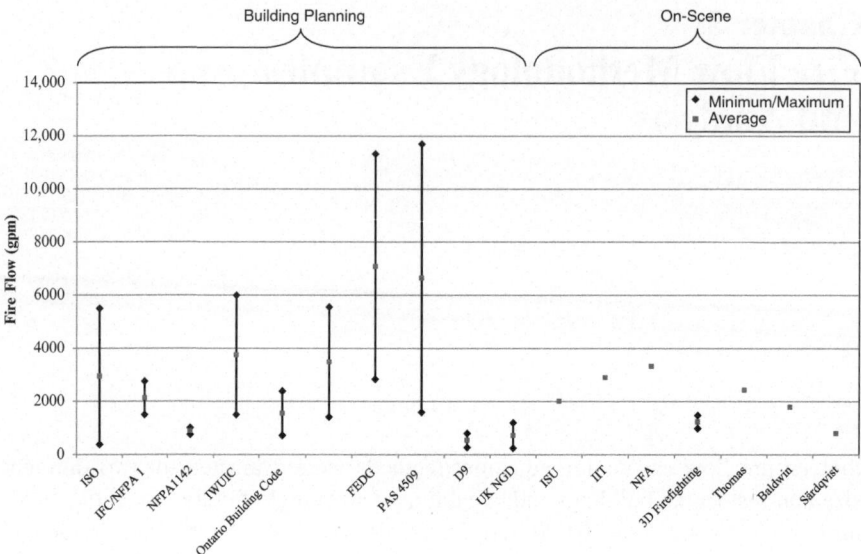

Fig. 3.1 Maximum and minimum fire flow calculations for a non-sprinklered 10,000 ft^2 non-residential building

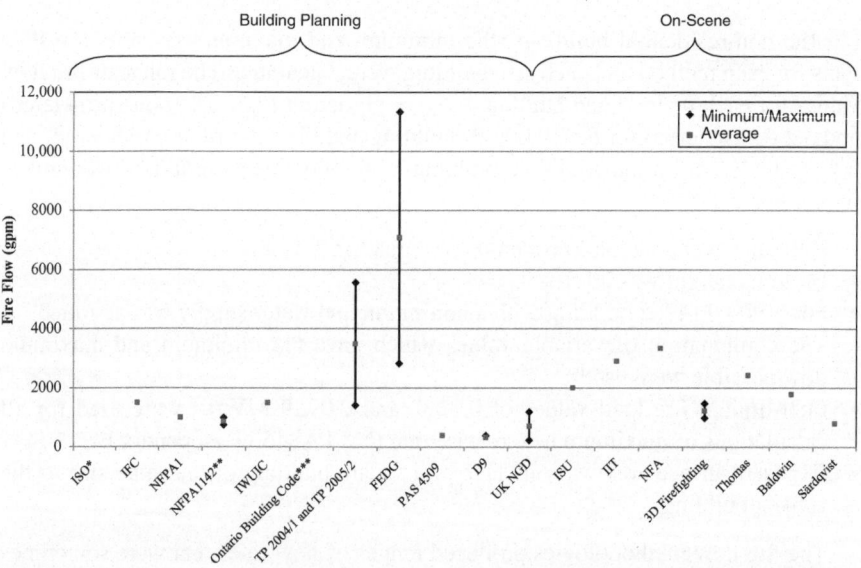

Fig. 3.2 Maximum and minimum fire flow calculations for a sprinklered 10,000 ft^2 non-residential building. *The ISO method does not calculate needed fire flows for sprinklered buildings. **The code does not specify the amount of reduction when sprinkler protection is provided. ***The fire flow is equal to the hose line demand required by NFPA 13

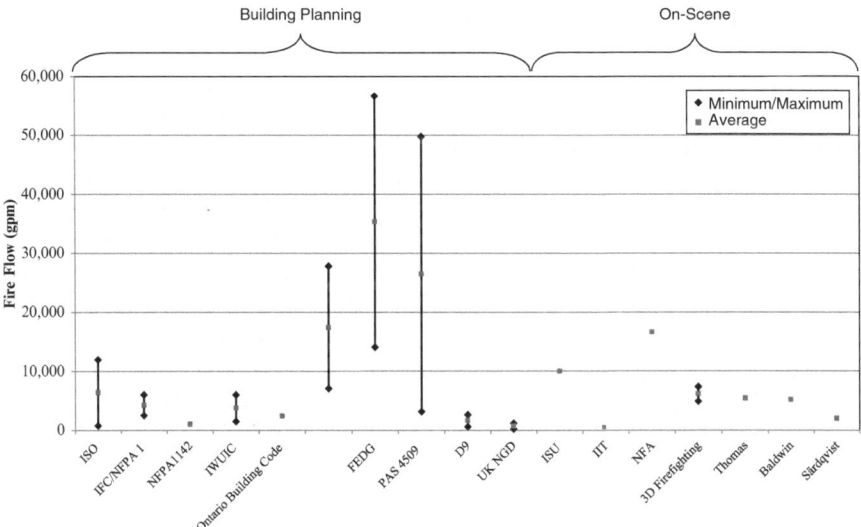

Fig. 3.3 Maximum and minimum fire flow calculations for a non-sprinklered, non-residential 50,000 ft^2 building. *Note*: IIT equation produces negative fire flows for areas above ~32,000 ft^2; this formula produces a value of 128 gpm at 32,000 ft^2, which is shown

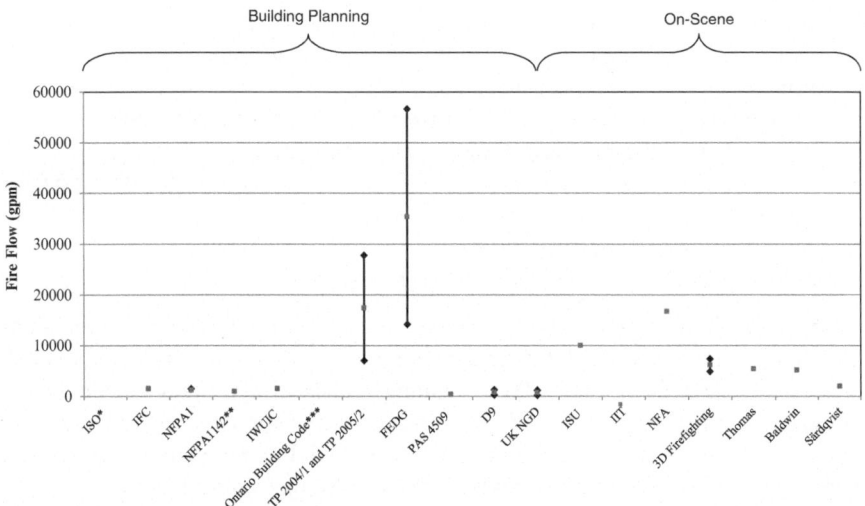

Fig. 3.4 Maximum and minimum fire flow calculations for a sprinklered, non-residential 50,000 ft^2 building. *Note*: IIT equation produces negative fire flows for areas above ~32,000 ft^2; this formula produces a value of 128 gpm at 32,000 ft^2, which is shown. *The ISO method does not calculate needed fire flows for sprinklered buildings. **The code does not specify the amount of reduction when sprinkler protection is provided. ***The fire flow is equal to the hose line demand required by NFPA 13

by other researchers (see Sect. 2.3 and Appendix A). The ISO, FEDG, PAS 4509, and TP2004/1 and TP 2005/2 methods produced the largest ranges of values for the same size non-sprinklered building. However, the range of fire flows required by the PAS4509 method were dramatically reduced when the building was provided with automatic sprinkler protection. The ISO method is not applicable to buildings with an approved automatic sprinkler system.

The building planning methods tended to provide fire flows are much higher than the on-scene methods for non-sprinklered, non-residential buildings (see Figs. 3.1 and 3.3). For sprinklered buildings, most of the fire flows were reduced to values less than the on-scene methods (see Figs. 3.2 and 3.4). Most of the on-scene methods do not account for sprinkler protection in their calculations. Given the wide range of variables present in the fire flow calculation methods, it is problematic to devise a single building that can be used to compare all of the methods without introducing biases to one or several of the methods.

3.2   Residential Buildings

For the single-family residential buildings, the following assumptions were made:

- Building was composed of one fire area
- Building height was 20 ft
- The building had two floors
- The floor area of each floor was half of the total area
- For NFPA 1142, a building with a non-municipal water supply was assumed
- Sprinkler protection was considered as a factor
- The building was of combustible construction (Type V or equivalent)
- There were no exposure hazards (adjacent buildings were 51 ft away)
- Fuel-limited fire load value of 0.0232 MW/ft^2 were used for all calculations of maximum heat release rate (NZ PAS4509, Appendix E)
- Both floors are fully involved fires
- For the ISO method, the current rules were applied

For comparison purposes, both sprinklered and non-sprinklered cases were considered. The fire flows calculated for the residential buildings are shown in Figs. 3.5 and 3.6 for the 1500 ft^2 and 3500 ft^2 buildings, respectively. For 11 methods, the presence of sprinklers did not affect the fire flow. For the Ontario Building code method, the fire flow requirement for a sprinklered building was equal to the demand of the system; in this case the sprinkler demand was not estimated. Only four of the fire flow methods reduced the fire flow because of the presence of sprinklers; these reductions were greater than or equal to 33 %.

In general, the building planning methods produced fire flows that were much lower than the on-scene methods for the 3500 ft^2 building. However, for the 1500 ft^2 building, all of the fire flows were 1000 gpm or lower. The fire flows calculated using the building planning methods were the same for all but two methods, the

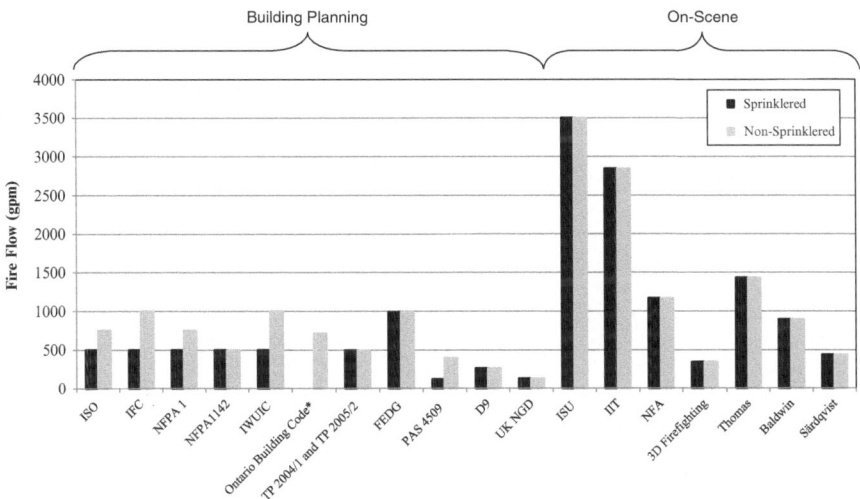

Fig. 3.5   Fire flow calculations for a 3500 ft² single-family home; sprinklered vs. non-sprinklered. *Fire flow is equal to the sprinkler demand

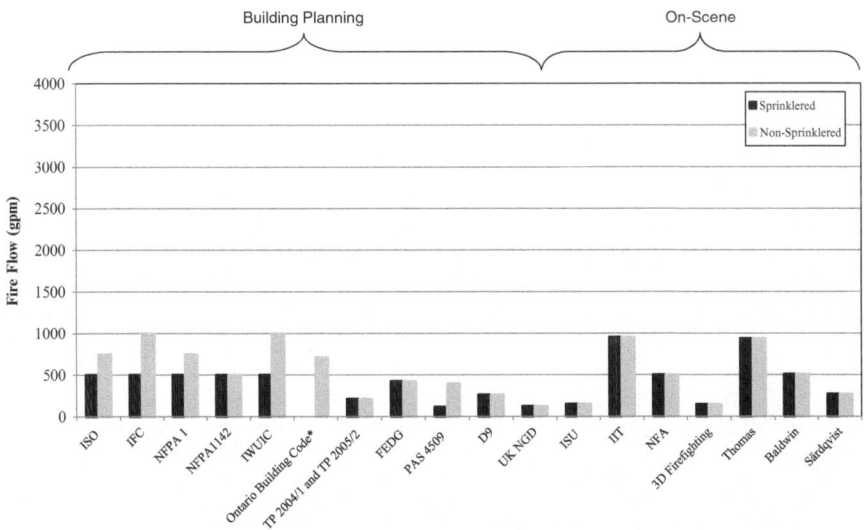

Fig. 3.6   Fire flow calculations for a 1500 ft² single-family home; sprinklered vs. non-sprinklered. *Fire flow is equal to the sprinkler demand

FEDG method and the TP 2004/1 and TP 2005/2 method which both calculate the fire flow based on heat release rate. The on-scene methods, with the exception of the IIT method, do not account for the occupancy of the building. However, the IIT method produced the second highest fire flow out of all of the methods for the 3500 ft² example.

Chapter 4
Gap Analysis

4.1 Important Variables

This GAP analysis attempts to identify the steps that are needed to determine which of the numerous methods (and associated variables) result in the most accurate fire flow assessment for modern buildings. The two categories of flow methodologies, building planning and on-scene, have differing application purposes. The focus of the GAP analysis is validating the building planning methods, since regulations for buildings/areas (i.e., what fire flow to require in fire codes and insurance guidelines) is of primary concern.

The variables are different for the two types of methods, as shown in the summary Table 4.1. Ultimately the methods are used to calculate the size of fire expected to be suppressed (and the fire flow required to do so) whether based on heat release rate or building area/volume. The on-scene methods use simple, scientifically-based formulas to derive the flow based on fire area or volume. But other metrics are needed to establish the reasonable fire size to be expected. These are embodied in the building planning methods.

The key variables in the building planning methods which are used de facto to derive fire size are the building geometry, and/or the rate of heat release/fire development which can be expected. The geometry is established literally: total height, number of stories, story height, length, width, area per floor, and total building area. Occupancy and building construction are de facto parameters to establish fire growth potential. Exposures would seem to be important; a separate handline or device(s) might be needed to prevent fire spread. Installed suppression systems, particularly sprinklers, should be considered as a potential fire-flow reduction parameter (with safety factor).

Utiskul and Wu [41] analyzed the effect of fire department total response time on the total amount of water used at a fire scene as summarized in Table 4.2. They defined the total response time as the time between fire department notification and the time at which suppression activities were initiated. In general, the longer the total

© Fire Protection Research Foundation 2015
M.E. Benfer, J.L. Scheffey, *Evaluation of Fire Flow Methodologies*,
SpringerBriefs in Fire, DOI 10.1007/978-1-4939-2889-7_4

Table 4.1 Comparison of the fire flow calculation methods and important factors used in the calculations

Method	Flow, duration, and/or volume	Empirical (loss data) or theoretical basis	Explicitly addresses residential	Building construction	Building occupancy and use	Exposure protection	Building or fire area	Building volume (height)	Heat release rate	Efficiency of water application or cooling	Human intervention	Reduction for sprinkler protection
Building planning methods												
ISO method	Flow, Duration[a]	Empirical	Yes	Yes	Yes	Yes	Yes	–	–	–	–	Yes
IFC/NFPA 1 methods	Flow, Duration	Empirical	Yes	Yes	–	–	Yes	–	–	–	–	Yes
NFPA1142	Flow, Volume	Unknown	–	Yes	Yes	Yes	–	Yes	–	–	–	Yes
IWUIC method	Flow, Duration	Unknown	Yes	–	Yes	–	Yes	–	–	–	–	Yes
Ontario building code method	Flow, Duration, Volume	Unknown	–	Yes	Yes	Yes	–	Yes	–	–	–	Yes
FIERAsystem method	Flow, Duration	Unknown	–	–	Yes	Yes	Yes	Yes	Yes	Yes	–	–
TP 2004/1 and TP 2005/2 methods	Flow, Volume	Theoretical	–	–	–	Yes	Yes	–	Yes	Yes	–	–
FEDG method (NZ)	Flow	Theoretical	–	–	–	–	–	–	Yes	Yes	–	–
PAS 4509 method	Flow, Duration, Volume	Unknown	Yes	–	Yes	Yes	Yes	Yes	Yes	–	Yes	Yes

(continued)

Method	Flow/Duration	Type									
D9 method	Flow / Duration	Unknown	Yes	Yes	–	Yes	(Yes)	–	–	Yes	Yes
UK national guidance document on the provision of water for firefighting	Flow	Unknown	–	Yes	–	Yes	–	–	–	–	–
On-scene methods											
Thomas, Baldwin, and Särdqvist methods	Flow	Empirical	–	–	–	Yes	–	–	–	–	–
ISU method	Flow	Empirical	–	–	–	–	Yes	–	–	–	–
IIT method	Flow	Empirical	–	Yes	–	Yes	–	–	–	–	–
NFA method	Flow	Empirical	–	–	–	Yes	–	–	–	–	–
3D firefighting method	Flow	Empirical	–	–	–	Yes	–	–	–	–	–

[a] The revised FSRS, which is yet to be approved, includes fire flow durations

Table 4.2 The effect of total response time on the water used at a fire scene [41]

Total response time (min)	Number of events	Range of water used in suppression activities (gal)	Average (gal)
5–10	18	100–4500	1412
10–15	7	280–11,000	2522
>15	3	1750–41,000	19,417

response times the more water was used during suppression activities. The total response time is linked to the fire development; during the response time, the fire is allowed to grow, where automatic fire suppression is not present or not activated, and thus will require more water to extinguish. Even though the data from these authors is limited, it illustrates the importance of fire department response time on fire flow requirements which most of the calculation methodologies do not include.

Although fire department response time would have an effect on the fire flow required at a fire scene, this variable is difficult to independently assess. It is difficult to predict and utilize in a fire flow methodology, since the pre-notification fire size cannot readily be determined. Sprinkler protection, and to a lesser extent, perhaps automatic detection, could be used as a predictor of fire size (inclusion of the sprinkler factor is recommended, see Sect. 4.2). Fire department response time might be assessed in a gross way for a limited data set using two parameters as suggested by the Table 4.1 data: response time less than 15 min and greater than 15 min. Fire department response more than 15 min might yield a methodology penalty (increase) for fire flow. Fire flow methodologies used with rural and suburban buildings implicitly account for possible longer fire department response times due to the remoteness of some buildings. In addition, these methods already assume there is greater risk where there are no municipal water supplies. It is recommended that rural and remote suburban buildings be addressed separately when assessing fire department response time.

The task now becomes one of obtaining validation data. The National Fire Incident Reporting System (NFIRS) is the most logical database to gather data from. NFIRS is the world's largest annual database of fire incident information with all 50 US states and the District of Columbia reporting incident data. NFIRS collects data on over one million fires each year. The NFIRS reporting systems vary from state to state in terms of the specific data collected. For a basic structure fires, NFIRS would gather the following data which could be relevant to fire flow calculations [42]:

- Incident location
- Incident type
- Dates and times (i.e., response time)
- Actions taken
- Property use
- Property details
- On-site materials
- Ignition: area, source, material ignited, equipment involved

- Fire origin and spread
- Fire suppression factors
- Structure type
- Building height and main floor size
- Number of stories damaged by flame
- Detector information (presence, type, power, operation, and effectiveness)
- Automatic extinguishing system information (presence, type, operation, effectiveness, failure reason)

This list of data covers the majority of variables used to calculate fire flows using the methodologies outlined in this book. The fire suppression factors data includes identification of inadequate water supply, and the means of water supply, whether a municipal water supply and/or tankers were used. The key factor not addressed in NFIRS is the estimated fire flow. The closest data reportable is the water supply flow, intended to reflect the sustained water supply capacity available for a period of 1 h to apparatus responding on the first alarm (reported in gallons per minute (gpm)).

There are a large number of occupancies which can be used in NFIRS reporting. These occupancies, and likely other data as well, would need to be consolidated in order to be useful for fire flow analysis. While there is a large amount of pertinent data that can be gleaned from NFIRS, it alone cannot readily be used to provide validation data for fire flow calculation methodologies since fire flow per se is not reported. In addition, NFIRS reports are not necessarily filled out completely or accurately by the responsible persons; this type of error should be addressed when assessing any NFIRS data.

4.2   Incident Data

An attempt was made to obtain fire flow data from public research; this yielded a single result. Utiskul and Wu [41] conducted a survey of fire departments across the US to gather data on water usage during residential fire suppression activities. The purpose of this Fire Protection Research Foundation (FPRF) sponsored study [41] was to provide guidance to water utilities and local jurisdictions on water usage and water meter performance during residential sprinkler system actuation. This survey was conducted by polling 25 community fire departments over a period of 5 months. The authors gathered information on the estimated fire flow, fire flow duration, total water used, building characteristics, sprinkler systems, hydrant information, fire conditions, actions taken, and fire event timeline. In 42 fires incidents which occurred in 2010. The estimated fire flow used during suppression activities was documented by the responding fire department. All of the fires occurred in 1- and 2-family dwellings with up to three stories; fires ranged from those which were contained in the room of origin to ones which spread beyond the structure of origin.

A histogram of the fire flows used in these residential fires is shown in Fig. 4.1. All but three of the fires had fire flows which were 450 gpm or less; the remaining three fires had fire flows of 1000, 1250, and 1750 gpm. Assuming that the fires

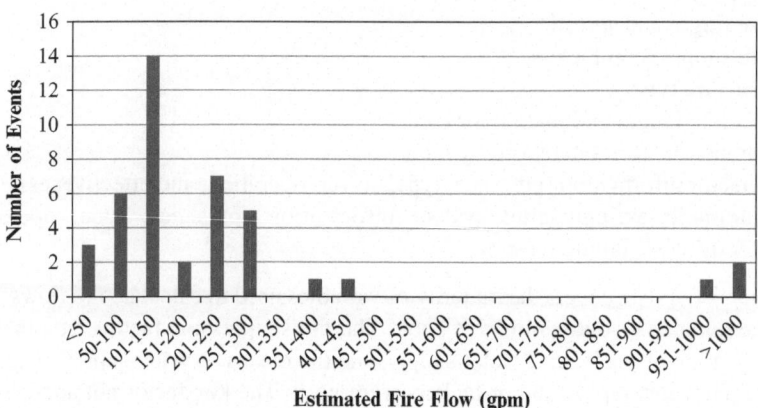

Fig. 4.1 Fire flow for residential fires from Utiskul and Wu [41]

occurred over a wide range of dwelling sizes (all of the dwelling sizes are not listed), the data suggests that a fire flow of approximately 1000 gpm (safety factor of 2.2) would be sufficient for most 1- and 2-family residential dwelling fires. This is consistent with the flow required using building planning fire flow calculation methodologies for residential dwellings of 3500 ft^2 or less (see Figs. 3.5 and 3.6); in most cases, less is permitted.

For 25 of the incidents reported by Utiskul and Wu, the size of the dwelling was able to be determined. This data was gathered specifically for this fire flow project by the NFPA Fire Analysis Services, using NFIRS data for the incidents (see Table 4.3). Fire flow rates for each of these dwellings were calculated using the ISO, NFPA 1, NFPA 1142, and NFA methods; shown in Table 4.3. Because the IFC and NFPA1 methods are essentially the same, only one method was evaluated. Fire flow durations and total required water volumes were calculated using the NFPA 1 and NFPA 1142 methods, as shown in Table 4.4. The fire flow duration and total volume is not specified by the ISO and NFA methods. The new ISO FSRS, which is yet to be approved, contains required fire flow durations for residential buildings. However, because this updated document is not yet approved or being used in practice, it was not evaluated for Table 4.4. Fire flow duration for the NFPA 1142 method was calculated by dividing the required water volume by the required fire flow.

In calculating the fire flow data for the 25 incidents, the following information and assumptions were used:

- Sprinklers were not present in the buildings evaluated
- Exposures were unknown; a range of values is given where applicable
- For the ISO method, the current rules (i.e., using the PPC3001 document) were applied
- Eight (8) foot high stories were assumed with the total area split evenly between floors for multi-story structures
- For the NFA formula, a fully involved fire for the entire structure was assumed
- For NFPA 1142, a building with a non-municipal water supply was assumed

Table 4.3 Calculated and estimated fire flow data for residential fires from Utiskul and Wu [41]

| Event | Total stories | Total area (ft²) | Est. water flow (gpm) | Building planning | | | On-scene |
				ISO (gpm)	NFPA 1 (gpm)	NFPA 1142 (gpm)	NFA (gpm)
1	2	3200	200	500–1500[a]	750–1000[a]	500	1067
2	1	1536	220	500–1500[a]	750–1000[a]	500	512
3	1	1500	300	500–1500[a]	750–1000[a]	500	500
4	1	1000	300	500–1500[a]	750–1000[a]	250–500[a]	333
5	3	10,125	400	3400	2750	750–1000[a]	3375
6	2	600	300	500–1500[a]	750–1000[a]	250–500[a]	200
7	1	320	30	500–1500[a]	750–1000[a]	250–500[a]	107
8	3	10,500	1750[b]	3612	2750	750–1000[a]	3500
9	1	1200	250	500–1500	750–1000[a]	250–500[a]	400
10	2	2200	110	500–1500	750–1000[a]	500	733
11	1	1294	120	500–1500	750–1000[a]	250–500[a]	431
12	1	1201	120	500–1500	750–1000[a]	250–500[a]	400
13	1	1053	150	500–1500	750–1000[a]	250–500[a]	351
14	2	3200	150	500–1500	750–1000[a]	500	1067
15	1	662	200	500–1500	750–1000[a]	250–500[a]	221
16	2	2156	1000	500–1500	750–1000[a]	500	719
17	1	1000	20	500–1500	750–1000[a]	250–500[a]	333
18	1	1000	100	500–1500	750–1000[a]	250–500[a]	333
19	1	2500	100	500–1500	750–1000[a]	500	833
20	1	1800	100	500–1500	750–1000[a]	500	600
21	1	1800	100	500–1500	750–1000[a]	500	600
23	1	600	125	500–1500	750–1000[a]	250–500[a]	200
24	2	5200	125	500–1500	2000	500–750[a]	1733
25	1	1100	125	500–1500	750–1000[a]	250–500[a]	367

[a]Specific flows can be determined if the separation distance is known
[b]750 gpm (hand lines for 48 min); 1000 gpm (deck gun for 5 min)

In general, the calculated fire flows for these 25 dwellings were greater than the estimated fire flow used during the actual fire event. This suggests that the fire flow methodologies are adequate for use with 1- and 2-family dwellings in that their requirements are rarely less than actual fire flows. For event number 8, the fire flow of 1750 gpm was significantly higher than that required by NFPA 1142. However, the NFPA 1142 requirement is the fire flow required within the first 5 min of arrival of the fire department. The 1750 gpm combined flow for this event consisted of the use of a deck-gun (i.e., water monitor) as well as hand-lines. The deck-gun was only used for 5 min of the suppression activities (at 1000 gpm), while hand lines were used for 48 min (at 750 gpm). There were no obvious correlations between the size of the building and the fire flow, fire flow duration, or total water volume used. It is unclear whether additional data would change this conclusion as the data from Utiskul and Wu covers a wide range of sizes of residential dwellings and a wide range of fire flows. In addition, the fire flow outliers, most of which are low outliers, tend to skew the correlations.

Table 4.4 Calculated and estimated fire flow duration and total water volume data for residential fires from Utiskul and Wu [41]

Event	Total stories	Total area (ft²)	Est. flow duration (min)	Est. total water used (gal)	NFPA 1 flow duration (min)	NFPA 1 total water (gal)	NFPA 1142 flow duration (min)	NFPA 1142 total water (gal)
1	2	3200	10	2000	60	45,000–60,000	10–16	5485–8228
2	1	1536	50	11,000	60	45,000–60,000	10–16	2633–3949
3	1	1500	10	3000	60	45,000–60,000	5–7	2571–3857
4	1	1000	13	3900	60	45,000–60,000	6–8	2000–3000
5	3	10,125	10	4000	120	330,000	23–26	17,357–26,035
6	2	600	15	4500	60	45,000–60,000	6–8	2000–3000
7	1	320	0.5	15	60	45,000–60,000	6–8	2000–3000
8	3	10,500	48[a]	41,000	120	330,000	24–27	18,000–27,000
9	1	1200	2	500	60	45,000–60,000	6–8	2057–3085
10	2	2200	2	220	60	45,000–60,000	7–11	3771–5657
11	1	1294	0.1	12	60	45,000–60,000	6–8	2218–3327
12	1	1201	0.15	18	60	45,000–60,000	6–8	2058–3088
13	1	1053	2	300	60	45,000–60,000	6–8	2000–3000
14	2	3200	0.5	75	60	45,000–60,000	10–16	5485–8228
15	1	662	2	400	60	45,000–60,000	6–8	2000–3000
16	2	2156	3	3000	60	45,000–60,000	7–11	3696–5544
17	1	1000	<1	<20	60	45,000–60,000	6–8	2000–3000
18	1	1000	0.5	50	60	45,000–60,000	6–8	2000–3000
19	1	2500	<0.5	<50	60	45,000–60,000	8–12	4285–6428
20	1	1800	2	200	60	45,000–60,000	6–9	3085–4628
21	1	1800	<0.5	<50	60	45,000–60,000	6–9	3085–4628
23	1	600	2	250	60	45,000–60,000	6–8	2000–3000
24	2	5200	3	375	120	240,000	17	8914–13,371
25	1	1100	2	250	60	45,000–60,000	6–8	2000–3000

[a]750 gpm (hand lines for 48 min); 1000 gpm (deck gun for 5 min)

The high range of the calculated fire flows are up to a factor of 75 (event #17) higher than the actual fire flows, which suggests that for many fires, flow methodologies may be conservative. However, the fire incident data set is very small. In addition, a safety factor should be considered. The only cases where the fire flow calculated was less than the actual fire flow was for event #4 (NFPA 1142 method), event #6 (NFPA 1142 and NFA methods), and event #16 (all methods). It is possible for incident #4 that the required fire flow calculated by the NFPA 1142 method could have been higher due to the additional water required for exposure protection. For incident #6, the small area of the house caused the fire flows calculated to be very low. For the NFPA 1142 fire flow, the calculation used from the standard is for suburban and rural areas where a municipal water supply is not present. This standard assumes more risk for the building (i.e., less water) than the other methods because these buildings tend to be more isolated. For event #16, it is possible that the flow calculation methodologies were inadequate. However, the actual duration of fire flow for this event was very short (3 min). This may have been a result of a very aggressive attack which put out the fire quickly. Alternately, the total fire flow (1000 gpm) may not have been necessary to suppress the fire.

The estimated fire flow durations for the fire incident data collected were much lower than those specified by NFPA 1. However, five of the events (#2, 3, 4, 6, and 8) had estimated flow durations which were greater than the calculated flow duration required by NFPA 1142. For all of these five events, the total water required by NFPA 1142 was also less than the estimated total water used at the scene, implying that the NFPA 1142 method was inadequate in these cases. However, this is not very surprising given that the NFPA 1142 method used was for a non-municipal water supply where large volumes of water may not always be present. The total water volume required by NFPA 1 was between 8 (event #8) and 640 (event #24) times greater than the estimated water used for flow durations of 1 min or longer. However, because the fire flow calculated for NFPA 1 is presumed to come from a municipal water supply (i.e., reductions are allowed by the AHJ for suburban/rural buildings), requiring such large volumes of water to be available over the course of an incident is not likely to be an issue.

Utiskul and Wu [41] analyzed the fire flow from residential sprinklers for a number of residences surveyed based on the system design and actual water supply present. Based on a single sprinkler operating, they determined that the expected flow discharged ranged from 22 to 38 gpm with an average of 28 gpm. Based on two sprinklers discharging, they determined that the expected flow discharged ranged from 26 to 55 gpm with an average of 39 gpm. Using a worst case approach, a sprinkler fire flow of 55 gpm would cover a range of systems and operating conditions. A similar conclusion of low expected fire flow was drawn by FM Global in an assessment of expected fire flow in sprinkler-protected areas [43]. These fire flows are significantly less than the calculated fire flows for residential buildings (see Figs. 3.5 and 3.6). The impact of sprinklers as a potential reduction factor appears to be appropriate and suggests that those methods which account for sprinkler protection may be inherently more appropriate.

4.3   Next Steps

The most logical approach to obtaining validation data would be to use a survey similar to the one used by Utiskul and Wu [41]. This type of survey yielded useful data and is relatively easy to analyze. Those departments which actually responded to this survey might be a good base to start with for a future fire flow data gathering effort, given their familiarity with a similar survey methodology. While NFIRS does provide a large amount of data which can be used in validating fire flow methodologies, it does not provide the fire flow itself. If NFIRS could be modified to include fire flow as a required parameter, this would be ideal given the large number of fires reported each year. This could provide a long-term data set for future validation of specific fire flow factors.

In the short term, a limited validation data set could be established by a survey similar to that performed by the FPRF for residential water usage. Since this data set will likely be sparse, all non-residential occupancies might be grouped into one or two general occupancies, say commercial and industrial buildings. Only data where a municipal water supply was present would be used in the validation study. This would largely eliminate rural incidents, which would be handled as special cases due to their remoteness. Mass conflagrations, explosions, etc., would not be included, since the study would be primarily concerned with typical or most likely situations to be encountered by a fire department and not necessarily the worst case scenario.

This survey data set could then be analyzed based on a variety of factors which might influence the fire flow including: response time, sprinkler presence and operation, building occupancy, building geometry, etc. NFIRS data on the incidents in the survey would supply supplementary information, besides that from the basic survey questionnaire regarding fire flow data. A possible survey form has been outlined in Appendix B; this survey form could also be made available to the fire departments online in order to facilitate easier data entry and management.

Chapter 5
Summary and Conclusions

There are a number of methods currently used to calculate required water flow rates for sprinklered and non-sprinklered properties. These methods are, in general, based on decades-old criteria derived using data from actual fires. Over the years, building construction methods, building contents, and fire suppression equipment and tactics have changed. The overall objective of this study was to assess the appropriateness of currently available fire flow methodologies. Nineteen existing fire flow calculation methods were identified and 16 were further examined. Three international fire flow methods were identified but could not be further examined due to the lack of a translated copy of the regulating document. Eleven methods evaluated were for the purpose of building planning. These were either from codified requirements or design guide documents. Five methods were on-scene methods for use by firefighters to evaluate a fire scene. Some of the methodologies evaluated also included requirements for determining the water supply volume and the required duration of the water supply.

In general, the building planning methods were more complicated and involved multiple steps and sub-calculations. These methods also typically accounted for many more variables (i.e., building construction, occupancy, fire size, etc.). The benefit of including more variables related to the building is that if the water supply for a given situation is found to be inadequate (i.e., water supply < minimum required fire flow), adjustments could be made to the building construction or protection features (e.g., adding a sprinkler system) to reduce the required fire flow. The building planning calculation methods also generally regulate other water supply features such as hydrant quantity and placement, water supply location, water supply duration, etc.

The on-scene fire flow calculation methods are much simpler. These methods consist of one equation with one variable, either the volume or area of the fire, making them easy and quick to use. This allows the firefighters on scene to assess whether they need more hose lines or apparatus to fight the fire. Other than as a first order approximation, the on-scene methods do not appear to lend themselves for use in codifying requirements. The building planning methods must be predictive; the on-scene methods are for real time situations where the potential volume of fire

© Fire Protection Research Foundation 2015
M.E. Benfer, J.L. Scheffey, *Evaluation of Fire Flow Methodologies*,
SpringerBriefs in Fire, DOI 10.1007/978-1-4939-2889-7_5

(magnitude) can be expertly judged by trained fire officers. Important variations in the building construction, use, or features such as exposure protection or the installation of sprinklers are de facto predictive parameters for establishing the magnitude of a fire which an arriving fire department should be able to handle.

Both the building planning and on-scene methods provide a large range of possible fire flows for a number of example/representative residential and non-residential buildings which were evaluated. These significant differences have been identified by other authors and indicate a need for additional field data to validate the methodologies.

The residential fire flow field data, while limited in number, was valuable in assessing the current predictive methods. Four of the existing fire flow calculation methods were compared to the field data. The fire flow methodologies were rarely less than the actual fire flows used. This suggests that, for residential occupancies, the fire flow calculations methodologies are conservative. A greater data set is necessary to completely validate assertion.

It appears that incorporating a sprinkler reduction factor in fire flow calculations is warranted. The fire flow used by the fire department for incidents where sprinklers operate is likely to be significantly less than if sprinklers were not present. This suggests that the methods which incorporate a sprinkler reduction factor may be more useful in establishing appropriate fire flow requirements.

This fire flow comparison was from a relatively small number of actual residential fires. More data is required to be able to sufficiently validate the fire flow methodologies over a range of occupancies. The key variables needed include actual fire flow used as a function of specific building geometry, exposures, and installed protection. The NFIRS reporting system is widely used by fire departments across the US to report fire incident data. All of the necessary data is included in the NFIRS report, except for an explicit estimate of fire flow used to suppress a given fire. The system does not currently contain a specific field for fire flow at the incident. If a fire flow field were added as a required item for NFIRS reporting, any fire department following the NFIRS format would provide validation data for the fire flow methodologies. This is a recommended long term approach.

In the near term, a survey of selected fire departments could be performed, similar to the approach used in the FPRF study of residential water supply [41]. The same departments used in that study could potentially be surveyed again. A possible survey form has been outlined (see Appendix B). Rural buildings would not be included in the initial analysis due to their remoteness and frequent use of non-municipal water supplies. For analysis of a small group of data, the occupancy of the building might be simply classified as either residential, commercial, or industrial. Depending on the extent of data gathered, refinement of the occupancy factor might be possible. This survey would not include major conflagrations or explosions which may skew the data. The data would be analyzed with respect to a number of factors including: building geometry, fire department response time, and sprinklers, which determine fire size. Exposures should be included since they may require a dedicated hose line. Investigation of the fire flow used for exposure protection could be expanded to include the effectiveness of that portion of fire flow in preventing fire spread to adjacent structures and reducing property loss.

Appendix A
Comparison of Fire Flow Calculations from Other Research

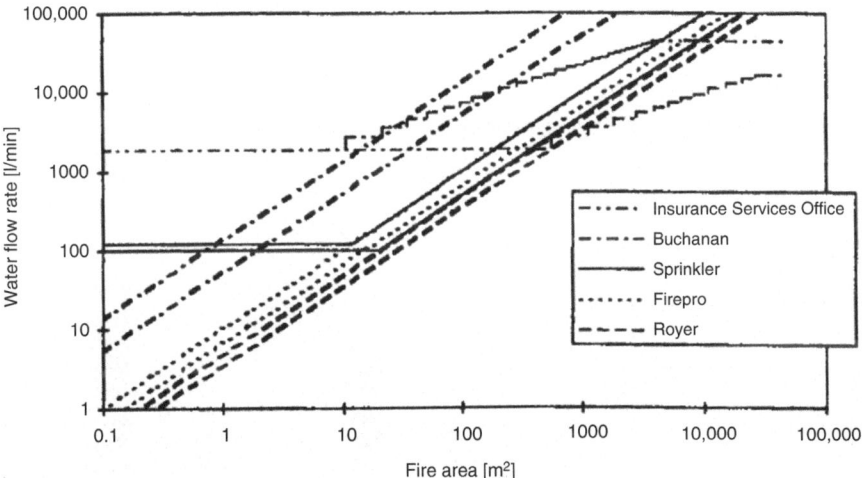

Fig. A1 Comparison of fire flow methodologies by Särdqvist [25]

© Fire Protection Research Foundation 2015
M.E. Benfer, J.L. Scheffey, *Evaluation of Fire Flow Methodologies*,
SpringerBriefs in Fire, DOI 10.1007/978-1-4939-2889-7

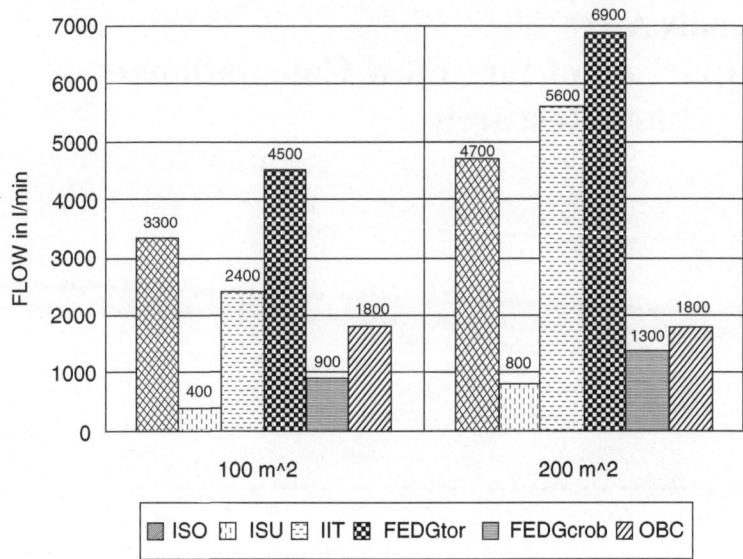

Fig. A2 Comparison of fire flow methodologies for residences by Barnett [38]

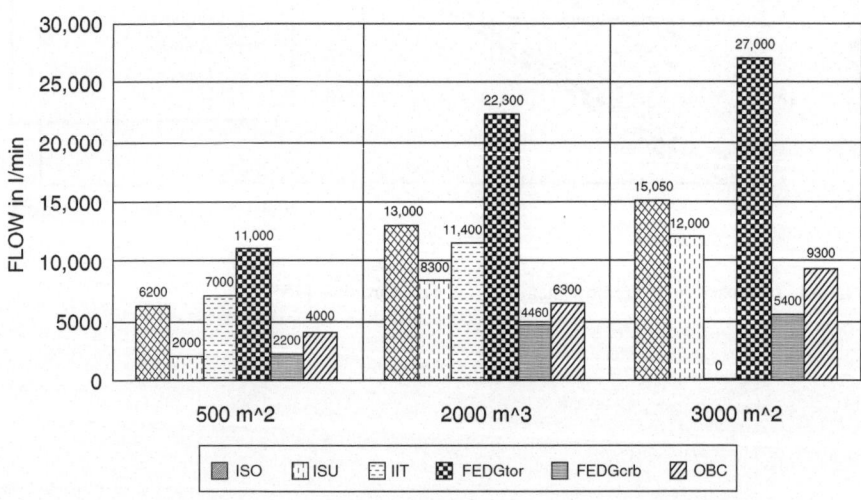

Fig. A3 Comparison of fire flow methodologies for offices by Barnett [38]

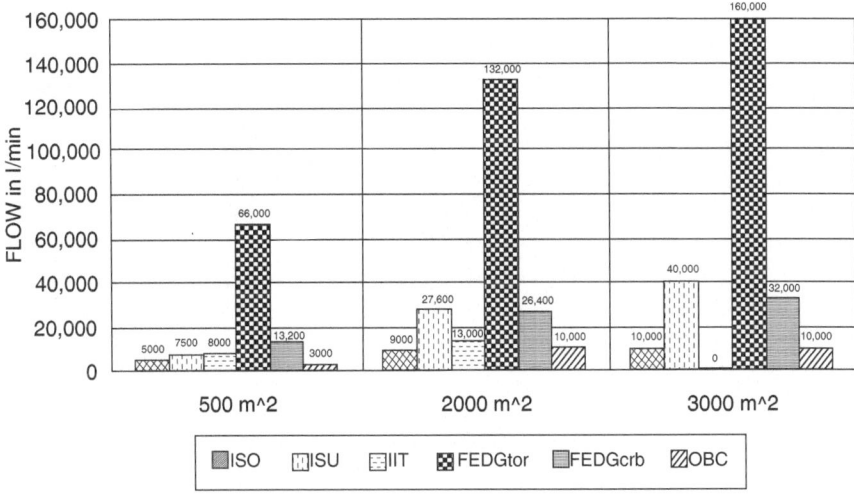

Fig. A4 Comparison of fire flow methodologies for warehouses by Barnett [38]

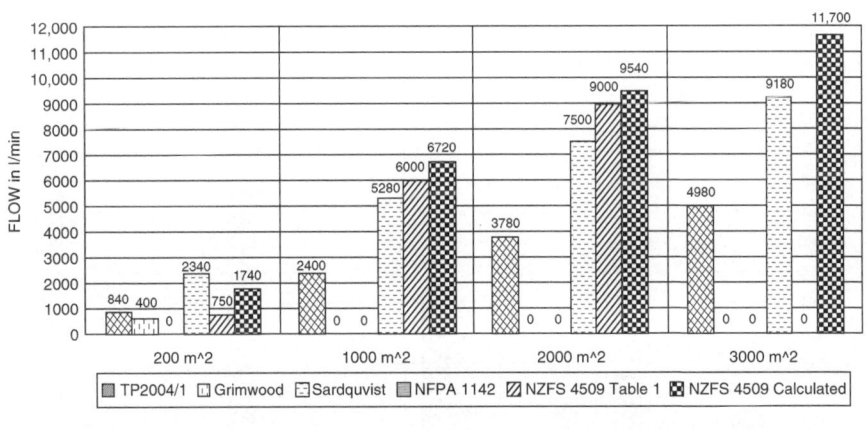

Notes
(1) TP2004/1 flows allows 50% for suppression and 50% for exposure.
(2) Grimwood flows based on 300 m^2 floor area maximum and 200 l/min/100m^2 for suppression only.
      No allowance made for exposure.
(3) NFPA 1142 flows are for suppression only. Exposure allowance is calculated separately with up
      to 25% extra for each building side.
      400 MJ/m^2 equates to NFPA Construction Classification 1.0 and Occupancy Hazard 7.
(4) NZFS 4509 "Table" flows are for suppression only. No allowance for exposure.
      Values are for Classes W3, W5, and W6 as in Table 2 of 4509.
(5) NZFS 4509 "Calc." flows are for Class W8 calculated using Appendices E and F of 4509.
      No allowance for exposure.

Fig. A5 Comparison of fire flow methodologies for buildings with FLED = 400 MJ/m^2 by Barnett [38]

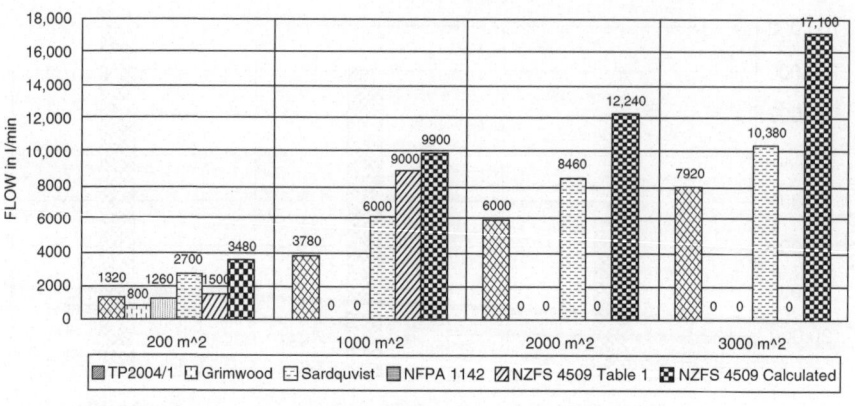

Notes

(1) TP2004/1 flows allows 50% for suppression and 50% for exposure.
(2) Grimwood flows based on 300 m^2 floor area maximum and 200 l/min/100m^2 for suppression only.
     No allowance is made for exposure.
(3) Sardqvist flows based on 200 m^ floor area. No allowance is made for expossure
(4) NFPA 1142 flows are for suppression only. Exposure allowance is calculated separately with up
     to 25% extra for each building side.
     400 MJ/m^2 equates to NFPA Construction Classification 1.0 and Occupancy Hazard 7.
(5) NZFS 4509 "Table" flows are for suppression only. No allowance for exposure.
     Values are for Classes W3, W5 and W6 as in Table 2 of 4509.
(6) NZFS 4509 "Calc." flows are for Class W8 calculated using Appendices E and F of 4509.
     No allowance is made for exposure.

Fig. A6 Comparison of fire flow methodologies for buildings with FLED = 800 MJ/m² by Barnett [38]

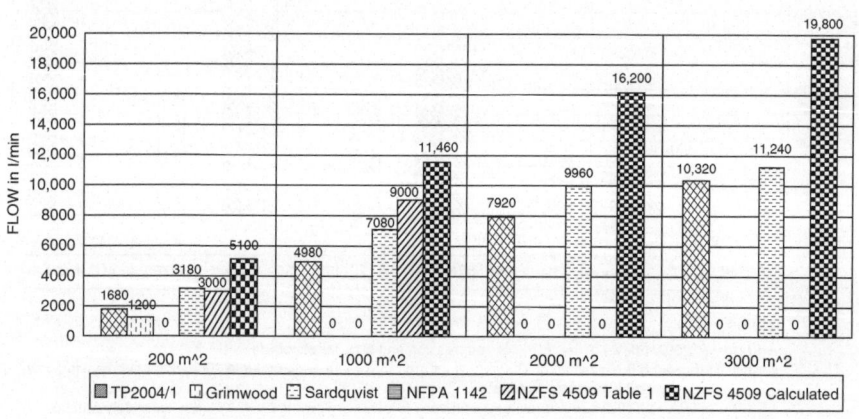

Notes

(1) TP2004/1 flows allows 50% for suppression and 50% for exposure.
(2) Grimwood flows based on 300 m^2 floor area maximum and 200 l/min/100m^2 for suppression only.
     No allowance made for exposure.
(3) NFPA 1142 flows are for suppression only. Exposure allowance is calculated separately with up
     to 25% extra for each building side.
     400 MJ/m^2 equates to NFPA Construction Classification 1.0 and Occupancy Hazard 7.
(4) NZFS 4509 "Table" flows are for suppression only. No allowance for exposure.
     Values are for Classes W3, W5, and W6 as in Table 2 of 4509.
(5) NZFS 4509 "Calc." flows are for Class W8 calculated using Appendices E and F of 4509.
     No allowance for exposure.

Fig. A7 Comparison of fire flow methodologies for buildings with FLED = 1200 MJ/m² by Barnett [38]

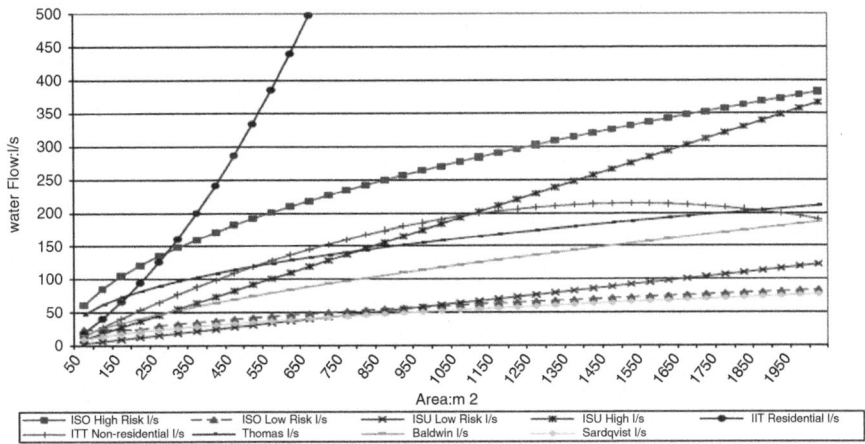

Fig. A8 Comparison of fire flow methodologies by Davis [39]

Appendix B
Example Form for Fire Flow Survey

Item	Units	Value	Notes
Fire department	–		
Incident number	–		
Type/description of fire	–		
Fire flow used	gpm		
Fire flow duration	min		
Total fire flow water used	gal		
Static pressure of water supply	psi		
Building area (total)	ft^2		
Building Dimensions	L×W×H (ft)		
Number of stories	–		
Building construction	–		
Fire department notification time	–		
Fire department arrival time	–		
Occupancy	–		
Exposures?	–		
Handline or device needed or used to protect exposure	–		
Sprinklers present?	–		
# of sprinklers operated?	–		
Sprinkler activation time	hh:mm		
Sprinkler shutdown time	hh:mm		
Sprinklers effective?	–		

© Fire Protection Research Foundation 2015
M.E. Benfer, J.L. Scheffey, *Evaluation of Fire Flow Methodologies*,
SpringerBriefs in Fire, DOI 10.1007/978-1-4939-2889-7

References

1. Wenzel, L.J., "Water Supply Requirements for Public Supply Systems," *NFPA Handbook, 20th Edition,* Section 15, Chapter 2, Quincy, MA.
2. AWWA M31, Distribution *System Requirements for Fire Protection,* American Water Works Association, Denver, CO, 1989.
3. *Guide for Determination of Needed Fire Flow,* Insurance Services Office, New Jersey, 2008.
4. Hutson, A. C., "Water Works Requirements for Fire Protection," *Journal of the American Water Works Association,* **40,** No. 9, 1948, p. 936.
5. *Fire Suppression Rating Schedule (Filed – Not Approved),* Insurance Services Office, New Jersey, 2012.
6. *International Fire Code 2012,* International Code Council, Illinois, 2011.
7. NFPA 1, *Fire Code 2012 Edition,* National Fire Protection Association, Quincy, MA, 2011.
8. *San Antonio, Texas, Code of Ordinances – Chapter 11, Article III, Section 11, Adoption of the 2012 International Fire Code,* San Antonio, TX.
9. NFPA 1142, *Water Supplies for Suburban and Rural Fire Fighting 2012 Edition,* National Fire Protection Association, Quincy, MA, 2011.
10. *International Wildland-Urban Interface Code 2012,* International Code Council, Illinois, 2011.
11. *Ontario Building Code 1997, July 1, 2005 Update,* Ministry of Municipal Affairs and Housing, Ontario, Canada, 2005.
12. NFPA 13, *Standard for the Installation of Sprinkler Systems 1996 Edition,* National Fire Protection Association, Quincy, MA,, 1996.
13. OFM-TG-03-1999, *Fire Protection Water Supply Guideline for Part 3 in the Ontario Building Code,* Ontario, Canada, 1999.
14. NFPA 1231, *Standard on Water Supplies for Suburban and Rural Fire Fighting,* National Fire Protection Association, Quincy, MA, 1975.
15. Torvi, D., Kashef, A., Benichou, N., Hadjisophocleous, G., "FIERAsystem Water Requirements Model (WTRM)," National Research Council of Canada, 2002.
16. "SFPE (NZ) Technical Publication - TP 2004/1, Calculation Methods for Water Flows Used For Fire Fighting Purposes," Society of Fire Protection Engineers (New Zealand), 2004.
17. "SFPE (NZ) Technical Publication - TP 2005/2, Calculation Methods for Storage Water Used For Fire Fighting Purposes," Society of Fire Protection Engineers (New Zealand), 2005.
18. Buchanan, A., H. (Editor), "New Zealand Fire Engineering Design Guide," University of Canterbury, New Zealand, 2001.
19. SNZ PAS 4509:2008, *New Zealand Fire Service Firefighting Water Supplies Code of Practice,* Standards New Zealand, 2008.

© Fire Protection Research Foundation 2015
M.E. Benfer, J.L. Scheffey, *Evaluation of Fire Flow Methodologies,*
SpringerBriefs in Fire, DOI 10.1007/978-1-4939-2889-7

20. D9 Technical Docume*nt, External Fire Control – Determination of Water Supply, INESC-FFSA-CNPP* Enterprise, France, 2001.
21. *National Guidance Document on the Provision of Water for Fire Fighting, Wa*ter UK and Local Government Association, 2007.
22. Iowa State University, Engineering Extension Service, Bulletin No. 18, "Water for Fire Fighting, Rate-of-Flow Formula," Iowa State University, Ames, IA, 1959.
23. Thomas, P.H., "Use of Water in the Extinction of Large Fires," *The Institution of Fire Engineers Quarterly* (19), no 35/1959, pp. 130–132.
24. Baldwin, R., "Use of water in the extinction of fires by brigades," *The Institution of Fire Engineers Quarterly* (31), no 82/1972, pp. 163–168.
25. Särdqvist, "Real Fire Data, Fires in non-residential premises in London 1994–1997," Department of Fire Safety Engineering, Lund University, Sweden, Report 7003, 1998.
26. Wieder, M.A., *Fire Service Hydraulics and Water Supply,* Fire Protection Publications, Oklahoma, 2005.
27. Grimwood, P., Hartin, E., McDonough, J., Raffel, S., *3D Firefighting – Training, Techniques, and Tactics,* Fire Protection Publications, Oklahoma, 2005.
28. PD 7974-5, *Application of Fire Safety Engineering Principles to the Design of Buildings Part 5 – Fire Service Intervention*, British Standards Institution, UK, 2002.
29. Shedd, J. H., Discussion on a paper by William B. Sherman, "Ratio of Pumping Capacity to Maximum Consumption," *Journal of New England Water Works Association,* **3**, 1889, p. 113.
30. Fanning, J. T., Distribution Mains and the Fire Service— *Proceedings of the American Water Works Association*, **12**, 1892, p. 61.
31. Kuichling, E., *The Financial Management of Water Works*—Transactions of the American Society of Civil Engineers, **38**, 1897, p. 16.
32. Freeman, J. R., "The Arrangement of Hydrants and Water Pipes for the Protection of a City against Fire," *Journal of the New England Water Works Association,* **7**, 1892, p. 49.
33. Metcalf, L., Kuichling, E., and Hawley, W. C., Some Fundamental Considerations in the Determination of a Reasonable Return for Public Fire Hydrant Service—*Proceedings of the American Water Works Association,* **31**, 1911, p. 55.
34. *Beheersbaarheid van Brand 2007,* Ministerie van BZK, Netherlands, 2007.
35. VdS 2034, *(Title translated: "Leaflet for the Evaluation of Non-Public Fire Departments"),* Verdag, Germany, 2003.
36. DVGW, *Arbeitsblatt W 405 (Subtitle translated: "Provision of Extinguishing Water Through The Public Drinking Water Supply"),*DVGW Technical, Germany, 2008.
37. Särdqvist, S., "Fire Brigade Use of Water," *Interflam '99*, Edinburgh, Scotland, 1999.
38. Barnett, C., "SFPE (NZ) Technical Publication - TP 2007/1, Comparison of 11 Methods for Determining Water Used For Fire Fighting," Society of Fire Protection Engineers (New Zealand), 2007.
39. Davis, S., "Fire Fighting Water: A Review of Fire Fighting Water Requirements - A New Zealand Perspective," University of Canterbury, New Zealand, 2000.
40. *The Measuremnent of Building Firesafety, Workbook*, Firepro Institute Ltd, 1995.
41. Utiskul, Y., Wu, N.P., "Residential Fire Sprinklers – Water Usage and Water Meter Performance Study," The Fire Protection Research Foundation, Quincy, MA, 2011.
42. "NFIRS Version 5.0 Design Documentation, Specification Release 2013.1," Federal Emergency Management Agency, Washington, DC, January 2013.
43. Wieczorek, C.J., Ditch, B., and Bill, R.G., Environmental Impact of Automatic Fire Sprinklers, Technical Report, FM Global Research Division, 2010.